IS WORK GOOD FOR YOUR HEALTH AND WELL-BEING?

Gordon Waddell, CBE DSc MD FRCS
Centre for Psychosocial and Disability Research, Cardiff University, UK

A Kim Burton, PhD DO EurErg
Centre for Health and Social Care Research, University of Huddersfield, UK

The authors were commissioned by the Department for Work and Pensions to conduct this independent review of the scientific evidence. The authors are solely responsible for the scientific content and the views expressed which do not necessarily represent the official views of the Department for Work and Pensions, HM Government or The Stationery Office.

London: TSO

TSO

Published by TSO (The Stationery Office) and available from:

Online
www.tsoshop.co.uk

Mail, Telephone, Fax & E-mail
TSO
PO Box 29, Norwich, NR3 1GN
Telephone orders/General enquiries: 0870 600 5522
Fax orders: 0870 600 5533
E-mail: customer.services@tso.co.uk
Textphone 0870 240 3701

TSO Shops
123 Kingsway, London, WC2B 6PQ
020 7242 6393 Fax 020 7242 6394
68-69 Bull Street, Birmingham B4 6AD
0121 236 9696 Fax 0121 236 9699
9-21 Princess Street, Manchester M60 8AS
0161 834 7201 Fax 0161 833 0634
16 Arthur Street, Belfast BT1 4GD
028 9023 8451 Fax 028 9023 5401
18-19 High Street, Cardiff CF10 1PT
029 2039 5548 Fax 029 2038 4347
71 Lothian Road, Edinburgh EH3 9AZ
0870 606 5566 Fax 0870 606 5588

TSO Accredited Agents
(see Yellow Pages)

and through good booksellers

The information contained in this publication is believed to be correct at the time of manufacture. Whilst care has been taken to ensure that the information is accurate, the publisher can accept no responsibility for any errors or omissions or for changes to the details given.

A CIP catalogue record for this book is available from the British Library.

A Library of Congress CIP catalogue record has been applied for.

First published 2006

ISBN 0 11 703694 3
13 digit ISBN 978 0 11 703694 9

Printed in the United Kingdom by The Stationery Office

Contents

Acknowledgements

We are grateful to Keith Palmer and Christopher Prinz for their careful review of the final draft of the report.

We thank the following colleagues for their helpful ideas and comments, and for pointing us to useful material during the course of the project: Kristina Alexanderson, Robert Barth, Jo Bowen, Peter Donceel, Hege Eriksen, Simon Francis, David Fryer, Bob Grove, Bill Gunnyeon, Elizabeth Gyngell, Bob Hassett, Camilla Ihlebaek, Nick Kendall, Rachel Lee, Chris Main, Fehmidah Munir, Trang Nguyen, Nick Niven-Jenkins, David Randolph, Justine Schneider, David Snashall, Holger Ursin, Keith Wiley, Nerys Williams, and Peter Wright.

Finally, we thank Debbie McStrafick for archiving the data and providing administrative support.

Executive summary

BACKGROUND

Increasing employment and supporting people into work are key elements of the UK Government's public health and welfare reform agendas. There are economic, social and moral arguments that work is the most effective way to improve the well-being of individuals, their families and their communities. There is also growing awareness that (long-term) worklessness is harmful to physical and mental health, so the corollary might be assumed – that work is beneficial for health. However, that does not necessarily follow.

This review collates and evaluates the evidence on the question 'Is work good for your health and well-being?' This forms part of the evidence base for the *Health, Work and Well-Being Strategy* published in October 2005.

METHODS

This review approached the question from various directions and incorporated an enormous range of scientific evidence, of differing type and quality, from a variety of disciplines, methodologies, and literatures. It a) evaluated the scientific evidence on the relationship between work, health and well-being; and b) to do that, it also had to make sense of the complex set of issues around work and health. This required a combination of a) a 'best evidence synthesis' that offered the flexibility to tackle heterogeneous evidence and complex socio-medical issues, and b) a rigorous methodology for rating the strength of the scientific evidence.

The review focused on adults of working age and the common health problems that account for two-thirds of sickness absence and long-term incapacity (i.e. mild/moderate mental health, musculoskeletal and cardio-respiratory conditions).

FINDINGS

Work: The generally accepted theoretical framework about work and well-being is based on extensive background evidence:

- Employment is generally the most important means of obtaining adequate economic resources, which are essential for material well-being and full participation in today's society;

- Work meets important psychosocial needs in societies where employment is the norm;

- Work is central to individual identity, social roles and social status;

- Employment and socio-economic status are the main drivers of social gradients in physical and mental health and mortality;

- Various physical and psychosocial aspects of work can also be hazards and pose a risk to health.

Unemployment: Conversely, there is a strong association between worklessness and poor health. This may be partly a health selection effect, but it is also to a large extent cause and effect. There is strong evidence that unemployment is generally harmful to health, including:

- higher mortality;

- poorer general health, long-standing illness, limiting longstanding illness;

- poorer mental health, psychological distress, minor psychological/psychiatric morbidity;

- higher medical consultation, medication consumption and hospital admission rates.

Re-employment: There is strong evidence that re-employment leads to improved self-esteem, improved general and mental health, and reduced psychological distress and minor psychiatric morbidity. The magnitude of this improvement is more or less comparable to the adverse effects of job loss.

Work for sick and disabled people: There is a broad consensus across multiple disciplines, disability groups, employers, unions, insurers and all political parties, based on extensive clinical experience and on principles of fairness and social justice. When their health condition permits, sick and disabled people (particularly those with 'common health problems') should be encouraged and supported to remain in or to (re)-enter work as soon as possible because it:

- is therapeutic;

- helps to promote recovery and rehabilitation;

- leads to better health outcomes;

- minimises the harmful physical, mental and social effects of long-term sickness absence;

- reduces the risk of long-term incapacity;

- promotes full participation in society, independence and human rights;

- reduces poverty;

- improves quality of life and well-being.

Health after moving off social security benefits: Claimants who move off benefits and (re)-enter work generally experience improvements in income, socio-economic status, mental and general health, and well-being. Those who move off benefits but do not enter work are more likely to report deterioration in health and well-being.

Provisos: Although the balance of the evidence is that work is generally good for health and well-being, for most people, there are three major provisos:

1. These findings are about average or group effects and should apply to most people to a greater or lesser extent; however, a minority of people may experience contrary health effects from work(lessness);

2. Beneficial health effects depend on the nature and quality of work (though there is insufficient evidence to define the physical and psychosocial characteristics of jobs and workplaces that are 'good' for health);

3. The social context must be taken into account, particularly social gradients in health and regional deprivation.

CONCLUSION

There is a strong evidence base showing that work is generally good for physical and mental health and well-being. Worklessness is associated with poorer physical and mental health and well-being. Work can be therapeutic and can reverse the adverse health effects of unemployment. That is true for healthy people of working age, for many disabled people, for most people with common health problems and for social security beneficiaries. The provisos are that account must be taken of the nature and quality of work and its social context; jobs should be safe and accommodating. Overall, the beneficial effects of work outweigh the risks of work, and are greater than the harmful effects of long-term unemployment or prolonged sickness absence. Work **is** generally good for health and well-being.

Health, work and well-being

Health is fundamental to human well-being, whilst work is an integral part of modern life. Increasing employment and supporting people into work are key elements of the UK Government's public health and welfare agendas (DH 2004; DWP 2006; HM Government 2005).

There are economic, social and moral arguments that, for those able to work, 'work is the best form of welfare' (Mead 1997; Deacon 1997; King & Wickam-Jones 1999) and is the most effective way to improve the well-being of these individuals, their families and their communities. There is also growing awareness that (long-term) worklessness is harmful to physical and mental health, so it could be assumed the corollary must be true – that work is beneficial for health. However, that does not necessarily follow. Therefore, the basic aim of this review is to consider the scientific evidence on the question 'Is work good for your health and well-being?'

This seemingly simple question must be placed in context. There are a number of potential causal pathways between health, work and well-being, with complex interactions and sometimes contradictory effects (Schwefel 1986; Shortt 1996):

- In modern society, work provides the material wherewithal for life and well-being

- Health and fitness underpin capacity for work (irrespective of whether any health problem bears a causal relationship to work – possible confounding)
    People's health may make them more or less likely to seek or obtain work, influence their work performance, and influence whether or not they leave work temporarily or permanently – health selection and the healthy worker effect.

- Work can be beneficial for health and fitness

- Work can carry risks for physical and mental health
    Certain jobs may create ill-health. People in certain kinds of work may be unhealthy because of non-work factors – possible confounding.

- Sickness and disability can impact on capacity for work
    Presenteeism, sickness absence, long-term incapacity, ill-health retirement.

- Work can be therapeutic. Conversely, (temporary) absence from work can be therapeutic

- Worklessness can be detrimental to health and well-being.

- Physical and mental health are important elements of well-being

- Work can have positive or negative effects on well-being.

Traditional approaches to occupational health and safety view work as a potential hazard and emphasise the adverse effects of work on health, and of ill health on capacity for work. But it is essential to consider the beneficial as well as the harmful effects of work on health and well-being (Figure 1). What ultimately matters is the balance between the positive and negative effects of work and how that compares with worklessness.

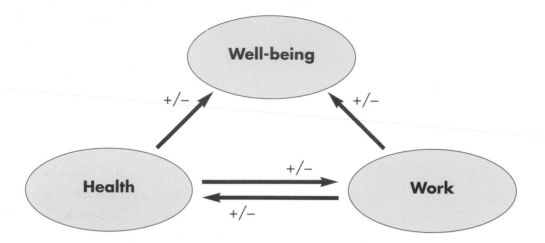

Figure 1. Possible causal pathways between health, work and well-being
(+/- : beneficial or harmful effects)

The main focus of this review is whether the current evidence suggests that work is (directly) beneficial for physical and mental health and well-being, and checking that any apparent relation is not explained by reverse causality or confounding. Whether or how work might cause (i.e. be a risk factor for) ill health is beyond the scope of this review because these are complex questions requiring different search strategies, in different literatures and with a different conceptual focus. However, that issue cannot be ignored when considering work and how it might affect the health of people with health problems (whatever their cause). This becomes important when advising people about continuing work or returning to work, in view of the concern that returning to (the same) work might do (further) harm. Associated questions include the timing of return to work and whether work demands should be modified. Therefore, key reviews of the epidemiological evidence about work as a risk factor are included and used to provide necessary balance when drawing up the evidence statements.

This review focuses on the 'common health problems' that now account for about two-thirds of sickness absence, long-term incapacity and early retirement - mild/moderate mental health, musculoskeletal and cardio-respiratory conditions (Waddell & Burton 2004). Many of these problems have high prevalence rates in the adult population, are essentially subjective, and often have limited evidence of objective disease or impairment. That is not to deny the reality of the symptoms or their impact, but these are essentially whole people, their health conditions are potentially remediable, and long-term incapacity is not inevitable. Moreover, epidemiology shows that these conditions are common whether in or out of work, risk factors are multifactorial, and cause-effect relationships ambiguous. Work, activity, and indeed life itself involves physical and mental effort, which imposes demands and is associated with bodily symptoms. Yet that effort is essential (physiologically) for maintaining health and capability.

More generally, the relationship between work and health must be placed in a broader social context. Account must be taken of the powerful social gradient in physical and mental health with socio-economic status – which is itself closely linked to work (Saunders 2002b; Saunders & Taylor 2002; McLean *et al.* 2005; Marmot & Wilkinson 2006). Social security covers diverse groups of people, with different kinds of problems, in very different circumstances. Many people receiving incapacity benefits have multiple disadvantages and face multiple barriers returning to work: older age, distance from the labour market, low skills, high local unemployment rates and employer discrimination (Waddell & Aylward 2005). Finally, social inequalities in work and health have a geographical dimension, with a strong link to deprived areas and local unemployment rates (McLean *et al.* 2005; Ritchie et al. 2005; Scottish Executive 2005). Analysis must be tempered by compassion for some of the most disadvantaged members of society, living in the most deprived circumstances (Rawls 1999; White 2004).

AIMS

This review considers the scientific evidence on the health effects of work and worklessness. It seeks the balance of the health benefits of work vs. the harmful effects of work, and of work vs. worklessness. It addresses the following questions:

1. Does the current evidence suggest that work is beneficial for physical and mental health and well being, in general and for common health problems?

2. What is the balance of benefits and risks to health from work and from worklessness?

3. Are there any circumstances (specific people, health conditions, or types of work) where work is likely to be detrimental to health and well-being?

4. Are there specific areas where there is a lack of evidence and need for further research?

DEFINITIONS

Analysis depends on understanding certain basic concepts. The following definitions will be used in this review, recognising that these and other concepts will require further debate and development as the *Health, Work and Well-being Strategy* evolves.

Work: involves the application of physical or mental effort, skills, knowledge or other personal resources, usually involves commitment over time, and has connotations of effort and a need to labour or exert oneself (Warr 1987; OECD 2003). Work is not only 'a job' or paid employment, but includes unpaid or voluntary work, education and training, family responsibilities and caring.

Worklessness: not engaged in any form of work, which includes but is broader than economic inactivity and unemployment.

Economic activity: covers all forms of engagement with the labour market, including: employed; self-employed; subsidised, supported or sheltered employment; and actively seeking work.

Economic inactivity: covers all those who are not engaged in the labour market, including those not actively seeking work, homemakers and carers, long-term sick and disabled, and retired (Barham 2002). There are now five times as many economically inactive as unemployed.

Employment: a job typically takes the form of a contractual relationship between the individual worker and an employer over time for financial (and other) remuneration, as a socially acceptable means of earning a living. It involves a specific set of technical and social tasks located within a certain physical and social context (Locke 1969; Warr 1987; Dodu 2005).

Unemployed: not employed at a job, wanting and available for work, and actively seeking employment (Barham 2002). This is often operationalised as being in receipt of unemployment benefits.

There is considerable overlap between 'health' and 'well-being', with philosophical debate about their relationship (Ryff & Singer 1998). Pragmatically, (Danna & Griffin 1999) suggest that *health* should be used when the focus is on the absence of physiological or psychological symptoms and morbidity; *well-being* should be used as a broader and more encompassing concept that takes account of 'the whole person' in their context.

Health: comprises physical and mental well-being, and (despite philosophical debate) is usually operationalised in terms of the absence of symptoms, illness and morbidity (WHO 1948; Danna & Griffin 1999; WHO 2004).

Well-being: is the subjective state of being healthy, happy, contented, comfortable and satisfied with one's quality of life. It includes physical, material, social, emotional ('happiness'), and development & activity dimensions (Felce & Perry 1995; Danna & Griffin 1999; Diener 2000).

Quality of life: is 'individuals' perception of their position in life in the context of the culture and value system in which they live and in relation to their own goals, expectations, standards and concerns' (The WHOQOL Group 1995).

Review Methods

This review had to do two things: (a) evaluate the scientific evidence on the relationship between work, health, and well-being; in order to do that, it also had to (b) make sense of, and impose some order on, the complex set of issues around work and health. It included a wide range of evidence, of differing type and quality, from a variety of disciplines, methodologies, and literatures. Meeting these diverse demands needed a combination of approaches. Developing concepts and organising the evidence required freedom to evolve as the project progressed. This may be described as a 'best evidence synthesis', which summarises the available literature and draws conclusions about the balance of evidence, based on its quality, quantity and consistency (Slavin 1995; Franche et al. 2005). This approach offered the flexibility needed to tackle heterogeneous evidence and complex socio-medical issues, together with quality assurance. At the same time, a rigorous approach was required when it came to assessing the strength of the scientific evidence.

The detailed methodology, including search strategies, inclusion/exclusion criteria, and evidence sources, is given in the Appendix. Throughout the review, broad and inclusive search strategies were used to retrieve as much material as possible, pertinent to the basic question: 'Is work good for your health and well-being?' Exclusion was primarily on the basis of lack of relevance to that question. Existing literature reviews, mainly from 1990 through early 2006, were used as the primary material, as in previous similar projects (Waddell & Burton 2004; Burton et al. 2004). Greatest weight was given to systematic reviews, whilst narrative reviews were used mainly to expand upon relevant issues or develop concepts. Selection inevitably involved judgements about quality: all articles were considered independently by both reviewers, and any disagreements resolved by discussion. Only in the absence of suitable reviews on a key issue was a search made for original studies. The focus was on common health problems, so major trauma and serious disease were included only if the evidence was particularly illuminating. The review covered adults of working age (generally 16-65 years).

ORGANISATION OF THE EVIDENCE

The structure of this report follows that of the literature searching and the evidence retrieved (Box 1). The obvious and most accessible starting point was reviews of the adverse health effects of unemployment. However, most of that evidence actually compares unemployment with work, so it was logical to expand the search to include the health effects of work and of unemployment. The retrieved literature was mainly about young or middle working-age adults, so a further search was made for material on older workers. All of these reviews considered the health impact of loss of employment, but provided little evidence on re-employment. A specific search was therefore made for individual longitudinal studies on the health impact of re-employment.

It then became apparent that that evidence was all about the impact of work or unemployment on people who were healthy. An additional search (in a different, largely clinical literature) was therefore made for reviews about the impact of work for sick and disabled people. The generic material retrieved showed a broad consensus of opinion but provided very little actual scientific evidence, so separate searches were made for condition-specific reviews on the three main categories of common health problems: mental health, musculoskeletal, and cardio-respiratory conditions.

Finally, recognising that social security is a special context, a separate search was made for literature on the health impact of moving off benefits and re-entering work. There are few reviews in this area, so original studies were also included. Workers compensation studies were excluded because they are not readily generalisable.

Information from the included papers was summarised and inserted into evidence tables (Tables 1 to 7), in chronological order.

Box1. The key areas of the review and the related evidence tables	
Areas of review	*Table*
Health effects of work vs. unemployment	Table 1
Health impacts of re-employment	Table 2
Work for sick and disabled people	Table 3
The impact of work on people with mental health conditions	Table 4
The impact of work on people with musculoskeletal conditions	Table 5
The impact of work on people with cardio-respiratory conditions	Table 6
Health after moving off social security benefits	Table 7

EVIDENCE SYNTHESIS AND RATING

Building on the evidence tables and using an iterative process, evidence statements were developed, refined, and agreed in each key area. The strength of the scientific evidence supporting each statement was rated as in Box 2. Where appropriate, the text of the evidence statements was used to expand on the nature or limitations of the underlying evidence, and to offer any caveats or cautions. The strength of the evidence should be distinguished from the size of the effect: e.g. there may be strong evidence about a particular link between work and health, yet the effect may be small. Furthermore, a statistical association does not necessarily mean a causal relationship. Where possible, effect sizes and causality are noted in the text of the evidence statements.

Box 2. Evidence rating system used to rate the strength of the scientific evidence for the evidence statements

	Scientific Evidence	Definition
***	**Strong**	generally consistent findings provided by (systematic review(s) of) multiple scientific studies.
**	**Moderate**	generally consistent findings provided by (review(s) of) fewer and/or methodologically weaker scientific studies.
*	**Weak**	*Limited evidence* – provided by (review(s) of) a single scientific study,
Mixed or conflicting evidence – inconsistent findings provided by (review(s) of) multiple scientific studies.
0	**Non-scientific**	legislation; practical, social or ethical considerations; guidance; general consensus.

The evidence statements are grouped and numbered under the areas in Box 1, and for ease of future referrence they are identified by the initial letter(s) of the heading concerned. Where the evidence statements were insufficient to convey complex underlying ideas, important issues were discussed in narrative text. Finally, the entire material was progressively distilled into an evidence synthesis to reflect the overall balance of the evidence about work and health. This was used to develop a conceptual framework located in the context of healthy working lives.

Quality assurance was provided by peer review of a final draft by two internationally acknowledged experts. Their feedback was used to refine the evidence statements and the evidence synthesis for the final report.

Review Findings

HEALTH EFFECTS OF WORK AND UNEMPLOYMENT

Table 1 lays out the retrieved evidence on the health impact of work (Table 1a) and of unemployment (Table 1b). Table 1c includes additional material on older workers.

Work

Extensive studies and theoretical analyses of work and of unemployment, and comparisons between work and unemployment, support the basic concept that work is beneficial for health and well-being:

W1　***　Employment is generally the most important means of obtaining adequate economic resources, which are essential for material well-being and full participation in today's society

> Table 1a: (Shah & Marks 2004; Layard 2004; Coats & Max 2005)
> Table 1b: (Jahoda 1982; Brenner & Mooney 1983; Nordenmark & Strandh 1999;
> Saunders 2002b; Saunders & Taylor 2002)

W2　***　Work meets important psychosocial needs in societies where employment is the norm

> Table 1a: (Dodu 2005), Table 1b: (Jahoda 1982; Warr 1987)

W3　***　Work is central to individual identity, social roles and social status

> Table 1a: (Shah & Marks 2004)
> Table 1b: (Brenner & Mooney 1983; Ezzy 1993; Nordenmark & Strandh 1999)

W4　***　At the same time, various aspects of work can be a hazard and pose a risk to health

> Table 1a: (Coggon 1994; Snashall 2003; HSC 2002; HSC 2004)

Logically, then, the nature and quality of work is important for health (WHO 1995; HDA 2004; Cox *et al.* 2004; Shah & Marks 2004; Layard 2004; Dodu 2005; Coats & Max 2005)). (All references in the following sub-section are to Table 1a).

W5　⁰　For moral, social and legal reasons, work should be as safe as reasonably practicable

> (WHO 1995; HSC 2002; HSC 2004)

W6   ⁰   Pay should be sufficient (though there is no evidence on what is 'sufficient')
[and the multiple non-health-related factors that influence pay levels must also
be acknowledged].

(Dooley 2003; Layard 2004; Coats & Max 2005)

W7   ***   There is a powerful social gradient in physical and mental health and mortality,
which probably outweighs (and is confounded with) all other work characteristics
that influence health.

(Acheson et al. 1998; Fryers et al. 2003; Coats & Max 2005)

W8   ***   Job insecurity has an adverse effect on health.

(Ferrie 1999; Benavides et al. 2000; Quinlan et al. 2001; Sverke et al. 2002; Dooley 2003)

W9   *   There is conflicting evidence that long working hours (with no evidence for any
particular limit) and shift work have a weak negative effect (Harrington 1994a;
Sparks et al. 1997; van der Hulst 2003); limited evidence that flexible work
schedules have a weak positive effect (Baltes et al. 1999); and conflicting evidence
about any effect of compressed working weeks of 12-hour shifts (Smith et al.
1998; Baltes et al. 1999; Poissonnet & Véron 2000) on physical and mental health.

In summary, there is a strong theoretical case, supported by a great deal of background
evidence, that work and paid employment are generally beneficial for physical and mental
health and well-being. The major proviso is that that depends on the quality of the job and the
social context. Nevertheless, the available evidence is on representative jobs, whatever their
quality and defects, and shows that on average they are beneficial for health. Within reason, shift
patterns and hours of work probably do not have a major impact on health: what workers
choose and are happy with is more important.

Most of this evidence is on men. What evidence is available suggests that the benefits of work are
broadly comparable for women, though that must be placed in the context of other gender,
family and caring roles.

W10   ***   Paid employment generally has beneficial or neutral effects and, importantly,
has no significant adverse effects on the physical and mental health of women.

(Klumb & Lampert 2004)

Unemployment

This section lays out in logical order the evidence on the association between unemployment
and health, on the causal relationship, on possible mechanisms and on modifying influences.
(All references in this section are to Table 1b).

There is a strong, positive **association** between unemployment and:

U1　***　Increased rates of overall mortality, mortality from cardiovascular disease, lung cancer and suicide.

(Brenner & Mooney 1983; Platt 1984; Jin *et al.* 1995; Lynge 1997;
Mathers & Schofield 1998; Brenner 2002)

U2　**　Poorer physical health (Mathers & Schofield 1998): e.g. cardiovascular risk factors such as hypertension and serum cholesterol (Jin *et al.* 1995), and susceptibility to respiratory infections (Cohen 1999).

U3　***　Poorer general health, somatic complaints, long-standing illness, limiting longstanding illness, disability [though these self-reported measures of health also correlate with psychological well-being].

(Jin *et al.* 1995; Shortt 1996; Mathers & Schofield 1998; Lakey 2001)

U4　***　Poorer mental health and psychological well-being, more psychological distress, minor psychological/psychiatric morbidity, increased rates of parasuicide.

(Platt 1984; Murphy & Athanasou 1999; Fryers *et al.* 2003)

U5　**　Higher medical consultation, medication consumption and hospital admission rates.

(Hammarström 1994b; Jin *et al.* 1995; Mathers & Schofield 1998; Lakey 2001)

Furthermore:

U6　***　There is strong evidence that unemployment can **cause, contribute to or aggravate** most of these adverse health outcomes.

(Bartley 1994; Janlert 1997; Shortt 1996; Murphy & Athanasou 1999)

There are a number of possible **mechanisms** by which unemployment might have adverse effects on health (Bartley 1994; Shortt 1996):

U7　***　The health effects of unemployment are at least partly mediated through socio-economic status, (probably relative rather than absolute) poverty and financial anxiety.

(Jahoda 1982; Brenner & Mooney 1983; Bartley 1994; Nordenmark & Strandh1999;
Saunders 2002b; Saunders & Taylor 2002; Brenner 2002; Fryers *et al.* 2003)

U8　⁰　Unemployment may affect physical health via a 'stress' pathway involving physiological changes such as hypertension and lowered immunity [though there is no direct evidence of this pathway in unemployed people].

(Ezzy 1993; Jin *et al.* 1995)

U9   ***   The psychosocial impact of being without a job can affect psychological health and lead to psychological/psychiatric morbidity.

(Jahoda 1982; Warr 1987; Ezzy 1993)

U10   *   There is conflicting evidence that unemployment is associated with altered health-related behaviour (e.g. smoking, alcohol, exercise).

(Bartley 1994; Hammarström 1994b; Jin *et al.* 1995)

U11   ***   One spell of unemployment may be followed by poorer subsequent employment patterns and increased risk of further spells of unemployment - the 'life course perspective'.

(Lakey 2001; McLean *et al.* 2005)

There is no clear evidence on the exact nature or relative importance of these causal mechanisms: any of them may play a part, and it appears likely they will vary in different individuals in different contexts for different outcomes (McLean *et al.* 2005; Bartley *et al.* 2005)

The impact of unemployment on health can be **modified** by:

U12   ***   socio-economic status, income and degree of financial anxiety.

(Hakim 1982; Brenner & Mooney 1983; Ezzy 1993; Bartley 1994; Shortt 1996; Cohen 1999; Nordenmark & Strandh 1999; Saunders 2002b; Saunders & Taylor 2002)

U13   ***   individual factors such as gender and family status, age, education, social capital, social support, previous job satisfaction & reason for job loss, duration out of work, and by desire and expectancy of re-employment.

(Warr 1987; Ezzy 1993; Hammarström 1994b; Banks 1995; Nordenmark & Strandh 1999; Lakey 2001; McLean *et al.* 2005)

U14   ***   regional deprivation and local unemployment rates.

(Brenner & Mooney 1983; McLean *et al.* 2005; Ritchie *et al.* 2005)

These factors may have positive, negative or sometimes quite complex effects on the health impact of unemployment. Moreover, it is not clear to what extent they a) have a direct impact on health, b) act as mediators, c) moderate the impact of unemployment, or d) act as confounders.

U15   ***   Despite the generally adverse effects of unemployment on health, for a minority of people (possibly 5-10%) unemployment can lead to improved health and well-being.

(Warr 1987; Ezzy 1993; Shortt 1996; Nordenmark & Strandh 1999)

Overall, there is extensive evidence that there are strong links between unemployment and poorer physical and mental health and mortality. A large part of this appears to be a cause-effect relationship, despite continuing debate about the relative importance of possible mechanisms. However, these adverse effects may vary in nature and degree for different individuals in different social contexts. Not all unemployment is 'bad': for a minority of people unemployment may be better for their health than their previous work. Nor does unemployment necessarily mean worklessness. Just as with work, health impacts depend on the quality of worklessness.

Age-specific findings

Both on a *priori* grounds and in the available evidence, three broad age groups can be distinguished: school leavers and young adults (16 to ~25 years); middle working age (~25 to ~50 years); and older workers (>50 years to retirement age).

School leavers and young adults:

Work and unemployment have different financial, social and health consequences for school leavers and young adults. They are at the start of their working lives, entering work for the first time, likely to have lesser financial and social commitments, and often still receiving some degree of parental family support. The majority are likely to be healthier, and health selection effects are therefore likely to be less important. (All references in this sub-section are to Table 1b)

A1   ***   The mortality rate of unemployed young people is significantly higher (compared with employed young people), mainly due to accidents and suicide.

(Hammarström 1994b; Morrell *et al.* 1998; Lakey 2001)

A2   *   There is mixed evidence that unemployment is harmful to the physical health of young people though any effect appears to be less than in middle working age or older workers.

(Hammarström 1994b; Morrell *et al.* 1998; Lakey 2001)

A3   ***   Unemployment has adverse effects on the mental health of young people (poor mental health and psychological well-being, more psychological distress, minor psychological/psychiatric morbidity) but these effects are generally less severe than in middle working age adults.

(Warr 1987; Hammarström 1994b; Morrell *et al.* 1998; Lakey 2001)

A4   *   There is mixed evidence that unemployed young people show worse health behaviour (compared with employed young people) on various measures e.g. eating habits, personal hygiene, sleeping habits, physical activities, alcohol, drugs and smoking.

(Hammarström 1994b; Morrell et al. 1998; Lakey 2001)

A5   *   There is mixed evidence that young unemployed people suffer adverse social consequences including social exclusion and alienation, financial deprivation, criminality and longer-lasting effects on employment patterns (including higher risk of further spells of unemployment) and health into adult life.

(Warr 1987; Hammarström 1994b; Lakey 2001)

A6   **   Young people from disadvantaged backgrounds, those with lower levels of education, or those who lack social support (characteristics which cluster together) are more vulnerable to the adverse health effects of unemployment.

(Hammarström 1994b; Lakey 2001)

Failure to enter the world of work and unemployment undoubtedly causes adverse effects on the physical and mental health and social well-being of school-leavers. However, the strength of these effects appears to be less than in adults, perhaps because of the resilience of youth and because their work habits are not yet established. For most, there appears to be relatively little impact on physical health, probably because they are healthier to start with. The impact on mental health is more comparable but still generally less than in adults, though that may depend on the young person's social context. The short-term social effects are again relatively mild, probably because of different social and family responsibilities, though the consequences of longer-term unemployment may be much more fundamental and important.

Middle working age

Most of this review and most of the available evidence is about middle working age adults, except where stated otherwise. (All references in this sub-section are to Table 1b).

All of the health effects of work and of unemployment are generally most marked in middle working-aged men, especially those with dependent families. (Hakim 1982; Warr 1987).

As with work, much of the evidence about unemployment is on men. Nevertheless, most of the available evidence suggests that the adverse health effects of unemployment are broadly comparable in men and women of middle working age, though they may be modified by gender and family roles. Single women with no family responsibilities may be more comparable to men. Women with partners and with family or caring commitments generally have less adverse health effects, possibly because they are financially cushioned and have better alternative social roles. (Warr 1987; Hammarström 1994b)

Older workers:

Work and unemployment have different financial, social and health consequences for older workers, particularly as they approach retirement. Early retirement may be a consequence of health problems (ill-health retirement), involuntary job loss (redundancy) or voluntary exit from the work force, each of which may have different financial, social and health effects. However, these patterns are often blurred (Aarts *et al.* 1996). There are methodological problems in separating the health impact of work, unemployment or retirement from that of ageing and from health selection effects (ill-health selection into retirement and the healthy worker effect). (The evidence statements for older workers are developed from both Table 1 and Table 2.)

(a) Work for older workers:

(References in this sub-section are to Table 1c).

A7 *** Physical and mental capability declines with age; thus work ability also declines but the nature and extent of the decline and the effect on work performance varies between individuals.

(Tuomi *et al.* 1997; Shephard 1999; Ilmarinen 2001; Benjamin & Wilson 2005)

A8 * There is mixed evidence that older workers have any decline in perceived/reported health (despite increasing disease prevalence).

(Tuomi *et al.* 1997; Wegman 1999; Shephard 1999; Scales & Scase 2000; Ilmarinen 2001)

A9 ** Older workers do not necessarily have substantially more sickness absence (despite more severe illnesses and injuries).

(Tuomi *et al.* 1997; Benjamin & Wilson 2005)

A10 ⁰ There is broad consensus that 1) 'work' should accommodate the needs and demands of ageing workers and 2) that physical ergonomics and work-organisational issues will contribute to safe participation in the workforce to older age.

(Hansson *et al.* 1997; Wegman 1999; Shephard 1999; Kilbom 1999; Ilmarinen 2001)

(b) The health impact of early retirement:

(References in this sub-section are to Tables 1 and 2.)

A11   ***   Early retirement can have either positive or negative effects on physical and mental health and mortality.

Table 1a: (Acheson *et al.* 1998), Table 1b: (Scales & Scase 2000)

Table 2c: (Ekerdt *et al.* 1983; Crowley 1986; Mein *et al.* 2003)

A12   **   Workers in lower and middle socioeconomic groups, those who are compulsorily retired or those who face economic insecurity in retirement (characteristics which cluster together) can experience detrimental effects on health and well-being and survival rates.

Table 1b: (Scales & Scase 2000), Table 2c: (Crowley 1986; Gallo *et al.* 2000; Gallo *et al.* 2001;

Gallo *et al.* 2004; Tsai *et al.* 2005)

A13   *   Workers in higher socio-economic groups, those who retire voluntarily or those who are economically secure in retirement (characteristics which cluster together) may experience beneficial effects on health and well-being

Table 1b: (Scales & Scase 2000), Table 2c: (Crowley 1986; Mein *et al.* 2003)

but there is some conflicting evidence.

(Morris *et al.* 1992).

A14   *   Early retirement out of unemployment may lead to improvement of the depression associated with unemployment.

Table 2c: (Frese 1987; Reitzes *et al.* 1996)

Demographic trends mean that older workers form an increasing proportion of the workforce. Some reduction in physical and mental capability and workability is probably inevitable with age, but chronological age is not a reliable marker. Many older workers are not only capable of continuing to work (Tsai *et al.* 2005) but want to do so (WHO 2001; AARP 2001).

There is a conceptual argument, and broad consensus, that matching work circumstances to the changing capabilities and needs of older workers will help to maintain their health and safety at work. That has yet to be tested, because most of the available evidence is from pragmatic studies of current practice without age-specific risk assessment or control. Nevertheless, it seems an entirely reasonable principle that would be simple and inexpensive to test.

The available evidence suggests that continuing to work, at least up to state retirement age, is not harmful to health or mortality in older workers (Gallo *et al.* 2004; Tsai *et al.* 2005; Pattani *et al.* 2004). This may, however, at least to some extent, reflect a health selection effect whereby those with more serious or chronic health problems leave the labour force. People who are happy with their current role (whether continuing to work or early retired) also have better affective well-being (Warr *et al.* 2004).

Conversely, early retirement can be either harmful or beneficial to physical and mental health and mortality, apparently depending largely on social determinants. Socio-economic group is not only a matter of financial and social status, but also reflects education, work type, social capital, lifestyle and behaviour. Other key determinants are (a) whether early retirement is by choice or involuntary and (b) financial (in)security in retirement: these tend to cluster with socio-economic group. Whether early retirement is good or bad for health appears to reflect powerful social gradients in health that continue after leaving work (Acheson *et al.* 1998; Scales & Scase 2000; Marmot 2004).

RE-EMPLOYMENT

The concept of re-employment for working age adults is relatively straightforward – moving from unemployment back into employment. In principle, it could also include moving from other forms of economic inactivity into employment, but this search did not retrieve any such studies. School leavers have not been employed before, so the closest equivalent is entering employment. Alternatively, they may move into some other form of 'work' such as further education or training. So, 're-employment' for school leavers was taken here to be any 'work' option other than unemployment. Older workers, just like working age adults, may be re-employed out of unemployment. Alternatively, they may move into (early) retirement, following which some may undertake other forms of 'work'. Unemployment and retirement may then have different effects on health and well-being and must be considered separately (Warr *et al.* 2004).

Table 2 presents the characteristics and key findings of the 53 retrieved longitudinal studies on the health impact of re-employment. The most common health outcomes were based on psychometrics, e.g. the General Health Questionnaire (GHQ), but a few studies gave clinical parameters such as blood pressure or mortality rates.

R1 ** Aggregate-level studies of employment rates show that increased employment rates lead to lower mortality rates.

Table 1b: (Brenner 2002)

School leavers and young adults

(References in this sub-section are to Table 2a unless stated otherwise.)

R2   ***   School leavers who move into employment or training, or return to education, show improvements in somatic and psychological symptoms compared with those who move into unemployment.

> (Banks & Jackson 1982; Donovan *et al.* 1986; Feather & O'Brien 1986; O'Brien & Feather 1990; Hammarström 1994a; Mean Patterson 1997; Bjarnason & Sigurdardottir 2003)

R3   ***   School leavers who move into 'unsatisfactory' employment can experience a decline in their health and well-being.

> (Patton & Noller 1984; Feather & O'Brien 1986; O'Brien & Feather 1990; Patton & Noller 1990; Hammarström 1994a; Dooley & Prause 1995; Schaufeli 1997)

R4   **   After re-employment, there is a persisting risk of subsequent poor employment patterns and further spells of unemployment.

> Table 1b: (Lakey 2001)

Adults

(References in this sub-section are to Table 2b unless stated otherwise.)

R5   ***   Re-employment of unemployed adults improves various measures of general health and well-being, such as self-esteem, self-rated health, self-satisfaction, physical health, financial concerns.

> (Cohn 1978; Payne & Jones 1987; Vinokur *et al.* 1987; Caplan *et al.* 1989; Kessler *et al.* 1989; Ferrie *et al.* 2001)

R6   ***   Re-employment of unemployed adults improves psychological distress and minor psychiatric morbidity.

> (Layton 1986b; Payne & Jones 1987; Iversen & Sabroe 1988; Kessler *et al.* 1989; Lahelma 1992; Hamilton *et al.* 1993; Claussen *et al.* 1993; Burchell 1994; Hamilton *et al.* 1997; Nordenmark & Strandh 1999; Liira & Leino-Arjas 1999; Vuori & Vesalainen 1999; Ferrie *et al.* 2001; Ferrie *et al.* 2002)

R7   ***   The beneficial effects of re-employment depend mainly on the security of the new job, and also on the individual's motivation, desires and satisfaction.

> (Kessler *et al.* 1989; Hamilton *et al.* 1993; Claussen *et al.* 1993; Burchell 1994; Wanberg 1995; Halvorsen 1998; Ferrie *et al.* 2001; Ferrie *et al.* 2002; Ostry *et al.* 2002)

R8   *   There is conflicting evidence that visits to health professionals are reduced by re-employment.

(Virtanen 1993; Ferrie *et al.* 2001)

R9   **   Even after re-employment, there is a persisting risk of subsequent poor employment patterns and further spells of unemployment.

Table 1b: (Saunders 2002b), Table 2b: (Liira & Leino-Arjas 1999)

Older workers

(References in this sub-section are to Table 2c.)

R10   **   Re-employment in older workers can improve physical functioning and mental health.

(Frese & Mohr 1987; Gallo *et al.* 2000; Pattani *et al.* 2004)

The studies in Table 2 provide strong evidence that re-employment leads to improved health in all age groups. However, the next question is whether that reflects cause and effect or could be explained by a health selection effect (the corollary of the healthy worker phenomenon). Three studies suggest that it is at least partly due to health selection (Hamilton *et al.* 1993; Claussen *et al.* 1993; Mean Patterson 1997). However, eight other studies that tested this hypothesis in various ways failed to demonstrate any health selection effect (Tiggemann & Winefield 1984; Warr & Jackson 1985; Layton 1986b; Kessler *et al.* 1989; Patton & Noller 1990; Graetz 1993; Schaufeli 1997; Vuori & Vesalainen 1999). Thus, the balance of the evidence is that health improvements are (at least to a large extent) a direct consequence of re-employment.

Moving into employment, continued education or training is clearly better than unemployment for the mental health, general well-being and longer-term social development of school leavers. That evidence is generally consistent but some studies show a smaller effect, perhaps reflecting different social and cultural contexts (e.g. (Patton & Noller 1990; Schaufeli 1997)). However, health benefits depend on the job or the training being 'satisfactory' while 'unsatisfactory' jobs may be little better than unemployment. That is consistent with Evidence Statements W5 – W9 about the importance of job quality.

In adults of middle working age, re-employment leads to clear benefits in psychological health and some measures of well-being, though there is a dearth of information on physical health. The magnitude of the improvement is more or less comparable to the adverse effects of job loss. The benefits of re-employment can be seen within the first year, and are generally sustained in those studies with a follow-up of some years.

Re-employment seems to have similar health benefits for older workers, but this is based on few studies. Moreover, the most important comparison may not be with continued unemployment but with (early) retirement, which can have either positive or negative effects on health (Evidence Statements A11 – A14). It is therefore not possible to predict which older workers will benefit from re-employment or under what circumstances, or whether re-employment will be better than other alternatives.

Re-employment generally leads to improved health, so efforts to seek a job are advisable. However, if these attempts to get work are unsuccessful, that failure can then have a further negative effect on mental health (Vinokur *et al.* 1987). Moreover, even if unemployed people do manage to get back to work, they remain at risk of further unemployment and subsequent poor employment patterns, which can have a longer-term impact on their health and well-being. Unemployment, like social disadvantage and deprivation, is best viewed across a life course perspective (Acheson *et al.* 1998; Bartley 1994).

WORK FOR SICK AND DISABLED PEOPLE

Table 3 shows a broad consensus across multiple disciplines and also, importantly, among disability groups, employers, unions, insurers, and the main political parties. It is widely accepted that job retention or (return to) work are desirable goals to maintain or improve quality of life and well-being. There is also general consensus that people should receive accurate, consistent information and advice, along with clinical and occupational management that reflects these goals (Coulter *et al.* 1998; Department of Health 2000; Detmer *et al.* 2003).

SD1   [0]   There is a broad consensus that, when possible, sick and disabled people should remain in work or return to work as soon as possible because it:

- is therapeutic;
- helps to promote recovery and rehabilitation;
- leads to better health outcomes;
- minimises the deleterious physical, mental and social effects of long-term sickness absence and worklessness;
- reduces the chances of chronic disability, long-term incapacity for work and social exclusion;
- promotes full participation in society, independence and human rights;
- reduces poverty;
- improves quality of life and well-being.

(Table 3)

The policy statements and guidance in Table 3 are based upon and reflect the available evidence, yet they are essentially expert opinions. Several refer to the evidence on the health benefits of work and the detrimental effects of unemployment in healthy people. Others discuss in general terms the harmful effects of prolonged sickness absence and avoidable incapacity, and the beneficial effects of work for sick people. However, there is little direct reference or linkage to scientific evidence on the physical or mental health benefits of (early) (return to) work for sick or disabled people.

MENTAL HEALTH

Table 4 presents the evidence on severe mental illness (Table 4a), common mental health problems (Table 4b) and 'stress' (Table 4c).

Severe mental illness

Severe mental illness was not the main focus of the present review but was included because it provides some of the best available evidence on work and mental illness. It may be argued that if work is good for people with severe mental illness that is likely to apply to a greater or lesser extent to people with mild/moderate problems. (References in this sub-section are to Table 4a).

M1   ***   Supported Employment programmes are effective for vocational outcomes in competitive employment (and more effective than Pre-Vocational Training).

(Crowther et al. 2001a; Bond 2004).

M2   **   Supported Employment, Pre-vocational Training and Sheltered Employment do not produce any significant effect (positive or negative) on health outcomes such as the psychiatric condition, severity of symptoms, or quality of life.

(Schneider 1998; Barton 1999; Crowther et al. 2001a; Schneider et al. 2002)

M3   **   There is a correlation between working and more positive outcomes in symptom levels, self-esteem, quality of life and social functioning, but a health selection effect is likely and a clear causal relationship has not been established.

(Schneider et al. 2002; Marwaha & Johnson 2004)

Many people with severe mental illness want to work and 30-50% are capable of work, though only 10-20% are working (Schneider 1998; Schneider et al. 2002; Marwaha & Johnson 2004). The current review shows that work is not harmful to the psychiatric condition or mental health of people with severe mental illness although, conversely, it has no direct beneficial impact on their mental condition either. However, the balance of the indirect evidence is that it is beneficial for their overall well-being (Schneider 1998; RCP 2002; Twamley et al. 2003).

Common mental health problems

(References in this sub-section are to Table 4b).

M4   ***   Emotional symptoms and minor psychological morbidity are very common in
the working age population: most people cope with these most of the time
without health care or sickness absence from work

(Ursin 1997; Glozier 2002)

M5   ***   People with mental health problems are more likely to be or to become workless
(sickness, disability, unemployment), with a risk of a downward spiral of
worklessness, deterioration in mental health and consequent reduced chances
of gaining employment.

(Merz et al. 2001; RCP 2002; Seymour & Grove 2005)

M6   ⁰   There is a general consensus that work is important in promoting mental health
and recovery from mental health problems and that losing one's job is detrimental.

(RCP 2002; Thomas et al. 2002; Seymour & Grove 2005)

There is limited evidence about the impact of (return to) work on (people with) mild/moderate
mental health problems, despite their epidemiological and social importance. However, there is
much more evidence on 'stress', which may be the best modern exemplar of common mental
health problems.

Stress

HSE defines stress as 'the adverse reaction people have to excessive pressure or other types of
demand placed on them' (HSE Stress homepage www.hse.gov.uk/stress accessed 24 January
2006). However, there are many other definitions of stress and no generally agreed scientific
definition (Wainwright & Calnan 2002; Palmore 2006). The term 'stress' is often used for both
psychosocial characteristics of work (stressors) and adverse health outcomes (stress responses).
To avoid fragmentation and duplication of the review, this section includes evidence on both
stressors and/or stress responses: these constructs should be distinguished. (References in this
sub-section are to Table 4c).

M7   ***   Cross-sectional studies show an association between various psychosocial
characteristics of work (job satisfaction, job demands/control, effort/reward,
social support) and various subjective measures of general health and
psychological well-being

(van der Doef & Maes 1999; Viswesvaran et al. 1999; de Lange et al. 2003; Tsutsumi & Kawakami 2004;
van Vegchel et al. 2005; Faragher et al. 2005)

The strongest associations are with job satisfaction (Faragher *et al.* 2005), and the weakest with social support (Viswesvaran *et al.* 1999; Bond *et al.* 2006). The associations are stronger for subjective perceptions of work than for more objective measures of work organization.

M8   ***   Longitudinal studies support a causal relationship between certain psychosocial characteristics of work (particularly demand and control) and mental health (mainly psychological distress) over time but the effect sizes are generally small.

(Viswesvaran *et al.* 1999; de Lange *et al.* 2003; Tsutsumi & Kawakami 2004; van Vegchel *et al.* 2005; Faragher *et al.* 2005; Bond *et al.* 2006)

The conceptual problem is the circularity in stimulus-response definitions: stressors are any (job) demands associated with adverse stress responses; stress responses are any adverse (health) effects attributed to stressors. The practical problem is that stressors and stress responses and the relationship between them are subjective perceptions, self-reported, open to modulation by the mental state identified as 'stress' (whatever its cause), and with confounding of cause and effect. There are no objective or agreed criteria for the definition or measurement of stressors or stress responses, or for the diagnosis of any clinical syndrome of 'stress' (Lazarus & Folkman 1984; Rick & Briner 2000; Rick *et al.* 2001; IIAC 2004; Wessely 2004). These conceptual and methodological problems create considerable uncertainty about psychosocial hazards, about psychosocial harms, and about the relationship between them (Rick & Briner 2000; Rick *et al.* 2002; Mackay *et al.* 2004; IIAC 2004; HSE/HSL 2005)

The underlying problem is the fundamental assumption that work demands/stressors are necessarily a hazard with potential adverse mental health consequences (Cox 1993; Cox *et al.* 2000a; Cox *et al.* 2000b; Mackay *et al.* 2004), ignoring or failing to take sufficient account of the possibility that work might also be good for mental health (Lazarus & Folkman 1984; Edwards & Cooper 1988; Payne 1999; Salovey *et al.* 2000; Briner 2000; Adisesh 2003; Nelson & Simmons 2003; Wessely 2004; HSE/HSL 2005; Dodu 2005). It is sometimes argued that this is a matter of quantitative exposure: 'Pressure is part and parcel of all work and helps to keep us motivated. But excessive pressure can lead to stress which undermines performance' (HSE Stress homepage **www.hse.gov.uk/stress** : accessed 24 January 2006). However, there is little evidence for such a dose-response relationship or for any threshold for adverse health effects (Rick & Briner 2000; Rick *et al.* 2001; Rick *et al.* 2002). Rather, work involves a complex set of psychosocial characteristics with which the worker interacts to experience beneficial and harmful effects on mental health. Other non-work-related issues can influence how the worker interacts with and copes with work stressors. Positive and negative work characteristics, positive and negative job-worker interactions, and positive and negative effects on the worker's health then all occur simultaneously. The final impact on the worker's health depends on the complex balance between them.

A more comprehensive model of mental health at work should embody the following principles:

- Safety at work should be distinguished from health and well-being. Safety is freedom from dangers or risks (Concise Oxford Dictionary). Health and well-being are much broader and more positive concepts.

- Personal perceptions, cognitions and emotions are central to the experience of 'stress' (Cox *et al.* 2000b; Rick *et al.* 2001; Rick *et al.* 2002; Ursin & Eriksen 2004).

- 'Stress' is both part of and reflects a wider process of interaction between the person (worker) and their (work) environment (Lazarus & Folkman 1984; Payne 1999; Cox *et al.* 2000b)

- Work can have both positive and negative effects on mental health and well-being (Lazarus & Folkman 1984; Edwards & Cooper 1988; Payne 1999; Briner 2000; Adisesh 2003; Nelson & Simmons 2003; HSE/HSL 2005)

This review did not retrieve any direct evidence on the relative balance of beneficial vs. harmful effects of work (of whatever psychosocial characteristics) on mental health and psychological well-being. Any adverse effects of work stressors appear to be comparable in magnitude to those of job insecurity (Ferrie 1999; Quinlan *et al.* 2001; Sverke *et al.* 2002). Any such effects are smaller than the adverse effects of unemployment (Jin *et al.* 1995; Mathers & Schofield 1998; Murphy & Athanasou 1999; Briner 2000; Glozier 2002), social gradients in health (Kaplan & Keil 1993; Acheson *et al.* 1998; Saunders 2002b) and regional deprivation (Saunders 2002b; Ritchie *et al.* 2005) on physical and mental health and mortality (Platt 1984; Lynge 1997; Mathers & Schofield 1998; Brenner 2002). There is no direct evidence on (a) how any adverse/beneficial effects of continuing to work compare with the adverse/beneficial effects of moving to sickness absence; (b) the balance of adverse or beneficial effects of return to work in people with stress-related health complaints; or (c) how any risk of adverse effects from returning to work compares with the adverse effects of prolonged sickness absence. On balance, any adverse effects of work on mental health appear to be outweighed by the beneficial effects of work on well-being and by the likely adverse effects of (long-term) sickness absence or unemployment.

MUSCULOSKELETAL CONDITIONS

Much of the literature retrieved on musculoskeletal conditions (Table A5) concerns low back pain, reflecting its occupational importance. However, many of the issues raised about back pain are common to other musculoskeletal conditions, particularly neck pain and arm pain (NIOSH 1997; Buckle & Devereux 1999; National Research Council 2001; Schonstein *et al.* 2003; National Health and Medical Research Council 2004; Helliwell & Taylor 2004; Waddell & Burton 2004; Punnett & Wegman 2004; Walker-Bone & Cooper 2005). (References in this section are to Table 5).

MS1　***　There is a high background prevalence of musculoskeletal conditions, yet most people with musculoskeletal conditions (including many with objective disease) can and do work, even when symptomatic.

(Burton 1997; De Beek & Hermans 2000; Waddell & Burton 2001; de Buck et al. 2002; Helliwell & Taylor 2004; de Croon et al. 2004; Walker-Bone & Cooper 2005; Henriksson et al. 2005; Burton et al. 2006)

MS2　***　Certain physical aspects of work are risk factors for the development of musculoskeletal symptoms and specific diseases. However, the effects sizes for physical factors alone are only modest, and tend to be confined to intense exposures.

(NIOSH 1997; National Research Council 1999; Buckle & Devereux 1999; Hoogendoorn et al. 1999; National Research Council 2001; Punnett & Wegman 2004; IIAC 2006)

MS3　***　Psychosocial factors (personal and occupational) exert a powerful effect on musculoskeletal symptoms and their consequences. They can act as obstacles to work retention and return to work; control of such obstacles can have a beneficial influence on outcomes such as pain, disability and sick leave.

(Burton 1997; Ferguson & Marras 1997; Davis & Heaney 2000; Abenhaim et al. 2000; National Research Council 2001; Waddell & Burton 2004; Helliwell & Taylor 2004; Woods 2005; Walker-Bone & Cooper 2005; Henriksson et al. 2005)

MS4　***　Activity-based rehabilitation and early return to work (or remaining at work) are therapeutic and beneficial for health and well-being for most workers with musculoskeletal conditions. [There is an underlying assumption that significant physical hazards should be controlled].

(Fordyce 1995; Frank et al. 1996; Abenhaim et al. 2000; de Buck et al. 2002; Staal et al. 2003; Carter & Birrell 2000; Schonstein et al. 2003; Waddell & Burton 2004; National Health and Medical Research Council 2004; COST B13 working group 2004; Helliwell & Taylor 2004; ARMA 2004; Staal et al. 2003; Cairns & Hotopf 2005)

MS5　**　Control (reduction) of the physical demands of work can facilitate work retention for people with musculoskeletal conditions, especially those with specific diseases.

(Frank et al. 1996; Westgaard & Winkel 1997; ACC and the National Health Committee 1997; Frank et al. 1998; RCGP 1999; de Buck et al. 2002; Staal et al. 2003; Waddell & Burton 2004; COST B13 working group 2004; Helliwell & Taylor 2004; de Croon et al. 2004; ARMA 2004; Franche et al. 2005; Loisel et al. 2005)

MS6   **   Organisational interventions, such as transitional work arrangements (temporary modified work) and improving communication between health care and the workplace, can facilitate early and sustained return to work.

(ACC and the National Health Committee 1997; Frank *et al.* 1998; Staal *et al.* 2003; Waddell & Burton 2004; COST B13 working group 2004; Henriksson *et al.* 2005; Franche *et al.* 2005; Loisel *et al.* 2005)

Four main themes emerged from the evidence: (a) the high background prevalence of musculoskeletal symptoms in the general population; (b) work can be a risk factor for musculoskeletal conditions; (c) the important modifying influence of psychosocial factors; and (d) the need to combine clinical and occupational strategies in the secondary prevention of chronic disability. Together, these themes are central to the relationship between work and health for people with musculoskeletal conditions.

The high background prevalence of musculoskeletal symptoms means that a substantial proportion of musculoskeletal conditions are not caused by work. Most people with musculoskeletal conditions continue to work; many patients with severe musculoskeletal diseases such as rheumatoid arthritis remain at work and experience health benefits (Fifield *et al.* 1991). Thus, musculoskeletal conditions do not *automatically* preclude physical work. Musculoskeletal symptoms (whatever their cause) may certainly make it harder to cope with physical demands at work, but that does not *necessarily* imply a causal relationship or indicate that work is causing (further) harm.

Biomechanical studies and epidemiological evidence show that high/intense exposures to physical demands at work can be risk factors for musculoskeletal symptoms, 'injury' and certain musculoskeletal conditions. However, causation is usually multifactorial and the scientific evidence is somewhat ambivalent: much depends on the outcome of interest. Physical demands at work can certainly precipitate or aggravate musculoskeletal symptoms and cause 'injuries' but, viewed overall, physical demands of work only account for a modest proportion of the impact of musculoskeletal symptoms in workers. The physical demands of modern work (assuming adequate risk control and except in very specific circumstances) play a modest role in the development of actual musculoskeletal pathology. In contrast, there is strong epidemiological and clinical evidence that (long-term) sickness absence and disability depend more on individual and work-related psychosocial factors than on biomedical factors or the physical demands of work (Walker-Bone & Cooper 2005).

More fundamentally, it is wrong to view physical demands from a purely negative perspective as 'hazards' with potential only to cause 'harm'. Different physical activities may either load or unload musculoskeletal structures. Physical activity is fundamental to physiological health and fitness and an essential part of rehabilitation from injury or illness. Work can be therapeutic. Thus, modern clinical management for most musculoskeletal conditions emphasises advice and support to remain in work or to return as soon as possible. People with musculoskeletal conditions who are helped to return to work can enjoy better health (level of pain, function, quality of life) than those who remain off work (Westman *et al.* 2006; Lötters *et al.* 2005). Importantly, physical activity and early return to work interventions do not seem to be associated with any increased risk of recurrences or further sickness absence (Staal *et al.* 2005; McCluskey *et al.* 2006).

The return to work process may need organisational interventions: risk reassessment and control, and modified work if required. The duration of modified work depends on the condition: for common musculoskeletal conditions such as back, neck or arm pain it should be temporary and transitional, although for chronic musculoskeletal disease such as rheumatoid arthritis it may be permanent. This approach is about accommodating the musculoskeletal condition (whatever its cause) rather than implying that work is causal or harmful.

CARDIO-RESPIRATORY CONDITIONS

Cardio-respiratory conditions can be severe and life-threatening yet, following appropriate treatment, recovery is often good with manageable residual impairment. Any persisting or recurring symptoms may then fit the description of a 'common health problem'. Cardio-respiratory conditions have a high prevalence in the general population (Perk & Alexanderson 2004; Tarlo & Liss 2005); whilst certain characteristics of work can be risk factors, cardio-respiratory conditions are often multifactorial in nature.

Table 6a presents the evidence on common cardiovascular conditions (myocardial infarction, heart failure and hypertension), arranged in two sections covering work as a risk factor (Table 6a-i) and management (Table 6a-ii). Table 6b presents the evidence on common respiratory conditions, particularly chronic obstructive pulmonary disease and asthma. Most of this literature was about the prevention, treatment or control of disease, rather than the impact of work on the health of people with cardio-respiratory conditions.

CR1   [0]   Returning workers with cardiovascular and respiratory conditions to work is a generally accepted goal that is incorporated into clinical guidance.

Table 6a: (Wenger *et al.* 1995; Thompson *et al.* 1996; van der Doef & Maes 1998; Thompson & Lewin 2000; Wozniak & Kittner 2002; Reynolds *et al.* 2004)

Table 6b: (Hyman 2005; Nicholson *et al.* 2005; HSE 2006)

CR2   ***   Many workers with cardiovascular and respiratory conditions do manage to return to work, but the rates vary and return to work may not be sustained.

Table 6a: (Shanfield 1990; Thompson et al. 1996; Dafoe & Cupper 1995; NHS CRD 1998; de Gaudemaris 2000; Wozniak & Kittner 2002; Perk & Alexanderson 2004)

Table 6b: (Malo 2005; Nicholson et al. 2005; Asthma UK 2004)

CR3   ⁰   The return to work process for workers with cardio-respiratory conditions is generally considered to require a combination of both clinical management and occupational risk control.

Table 6a: (Wenger et al. 1995; Dafoe & Cupper 1995)

Table 6b: (Hyman 2005; Nicholson et al. 2005)

CR4   *   There is limited evidence that rehabilitation and return to work for workers with cardio-respiratory conditions can be beneficial for general health and well-being and quality of life.

Table 6a: (Brezinka & Kittel 1995; Dafoe & Cupper 1995)

Table 6b: (Gibson et al. 2003; Lacasse et al. 2003; Hyman 2005)

CR5   ***   Prevention of further exposure is fundamental to the clinical management and rehabilitation of occupational asthma.

Table 6b: (Asthma UK 2004; Tarlo & Liss 2005; Malo 2005; IIAC 2006; HSE 2006)

There is an extensive literature on the rehabilitation of patients with cardiovascular conditions, though there is less on respiratory conditions. Workers who have experienced severe and potentially life-threatening illness face perceptual, work-related and social obstacles in returning to work, whether or not they have any continuing medical impairment. Nevertheless, of particular importance for the purpose of the present review, many of them can and do successfully return to work. Multimodal rehabilitation with control of workplace demands and exposures may facilitate that goal. However, there are significant difficulties in engaging patients in rehabilitation programmes (Newman 2004; Witt et al. 2005) which is partly a matter of service provision but also of motivation. There remains the issue of work retention, because patients often leave work again (Thompson et al. 1996; NHS CRD 1998).

The overall thrust of this literature is that return to (suitably controlled) work is an appropriate and desirable goal for many people with cardio-respiratory conditions. There is some evidence on the effectiveness of this approach for occupational outcomes, but there is little direct evidence about the impact of (return to) work on cardio-respiratory health. There is some indication that early return to work is safe for myocardial infarction patients stratified as low risk (Kovoor et al. 2006), and that patients with cardiopulmonary disease are rarely harmed by return to work recommendations (Hyman 2005). Furthermore, the limited evidence that is available suggests there may be some general health benefit (Brezinka & Kittel 1995) and this may extend to remaining in work (Gallo et al. 2004).

SOCIAL SECURITY STUDIES

There is a theoretical argument that moving off benefits and into work is likely to increase income, reduce poverty, increase human/social capital, and improve self-esteem and social status. In principle, that should move claimants up the social gradient in health, and thus improve their physical and mental health, quality of life and well-being (Acheson *et al.* 1998; Waddell & Aylward 2005). However, moving off benefits does not necessarily mean (re)-entering work, and the two must be distinguished. The further caveat is that any impact may depend on the nature and the quality of the job (Mowlam & Lewis 2005). (All references in this section are to Table 7).

SS1 *** Improvements in health and well-being from coming off benefits are associated with (re-)entering work, not simply with leaving the benefits system.

(Bound 1989; Caplan *et al.* 1989; Proudfoot *et al.* 1997; Dorsett *et al.* 1998; Watson *et al.* 2004;
Mowlam & Lewis 2005)

SS2 ** Claimants who move off benefits and (re-)enter work generally have increased income.

(Moylan *et al.* 1984; Caplan *et al.* 1989; Garman *et al.* 1992; Dorsett *et al.* 1998)

SS3 ** Moving off benefits and (re-)entering work is generally associated with improved psychological health and quality of life.

(Caplan *et al.* 1989; Erens & Ghate 1993; Vinokur *et al.* 1995; Rowlingson & Berthoud 1996;
Proudfoot *et al.* 1997; Dorsett *et al.* 1998; Watson *et al.* 2004; Mowlam & Lewis 2005)

There is conflicting evidence on the extent to which this is a health selection effect or cause and effect: probably both occur.

(Vinokur *et al.* 1995; Proudfoot *et al.* 1997; Bloch & Prins 2001; Watson *et al.* 2004)

SS4 *** After leaving benefits, many claimants go into poorly paid or low quality jobs, and insecure, unstable or unsustained employment. Many go on to further periods of unemployment or sickness, and further spell(s) on the same or other social security benefits.

(Daniel 1983; Ashworth *et al.* 2001; Hedges & Sykes 2001; Juvonen-Posti *et al.* 2002;
Bacon 2002; Bowling *et al.* 2004)

SS5 *** Claimants whose benefit claims are disallowed often do not return to work but cycle between different benefits and often report a deterioration in mental health, quality of life and well-being.

(Dorsett *et al.* 1998; Rosenheck *et al.* 2000; Ashworth *et al.* 2001)

Because the English-language literature in this area is mainly from the UK and the US, these conclusions relate to the social security systems in these countries. Moving off benefits might imply something different in other countries with different benefit systems and benefit levels.

Moving off benefits can have either positive or negative effects on health and well-being, depending mainly on how claimants leave benefits and whether or not they (re)-enter work. Of those claimants who leave benefits voluntarily, the majority (re)-enter work and have increased income, and many report that their health is completely recovered or much better. Of those claimants who are disallowed benefits, a minority (re)-enter work and their income generally falls, and many feel that their health remains unchanged or gets worse. Of those who are disallowed and appeal, very few (re)-enter work, and most feel that their health remains unchanged or gets worse. There are obvious (self)-selection effects in these divergent paths, which are also linked to social inequalities, multiple disadvantage and regional deprivation. The net result is that interventions which encourage and support claimants to come off benefits and successfully get them (back) into work are likely to improve their health and well-being; interventions which simply force claimants off benefits are more likely to harm their health and well-being (Dorsett *et al.* 1998; Ford *et al.* 2000; Rosenheck *et al.* 2000; Ashworth *et al.* 2001; Waddell 2004b; Waddell & Aylward 2005).

DISCUSSION

So, is work good for your health and well-being? This review found much more evidence than originally anticipated, even if it was of widely varying source, type and quality. Basically, there is a limited amount of high quality scientific evidence that directly addresses the question. However, there is a strong body of indirect evidence that can be built into a convincing answer: Yes, work is generally good for your health and well-being, with certain important provisos.

There is a generally accepted theoretical framework about work and well-being, based on extensive background evidence:

- Employment is generally the most important means of obtaining adequate economic resources, which are essential for material well-being and full participation in today's society;

- Work meets important psychosocial needs in societies where employment is the norm;

- Work is central to individual identity, social roles and social status;

- Employment and socio-economic status are the main drivers of social gradients in physical and mental health and mortality;

However, various physical and psychosocial aspects of work can also be hazards and pose a risk to health.

There is also a strong association between worklessness and poor health. Poor health can increase the risk of worklessness, whether that is in the form of disability, sickness absence or unemployment. Conversely, for many people, worklessness can have significant adverse effects on health. There may be an interaction between worklessness and poor health over time to produce a downward spiral of worklessness and health deterioration, which may be more marked for mental health problems. This review found strong evidence that unemployment is generally harmful to health, including:

- higher mortality;

- poorer general health, long-standing illness, limiting longstanding illness;

- poorer mental health, psychological distress, minor psychological/psychiatric morbidity;

- higher medical consultation, medication consumption and hospital admission rates.

The first comprehensive review of re-employment, presented here, provides strong scientific evidence that re-employment is associated with improved self-esteem, improved general and mental health, and reduced psychological distress and minor psychiatric morbidity. The magnitude of this improvement is more or less comparable to the adverse effects of job loss.

So, it may reasonably be concluded that work is better than unemployment for physical and mental health and well-being. That could be, to at least some extent, a health selection effect - those who stay in work or who (re)-enter work manage to do so because they are healthier; those who are less healthy may be more likely to become and remain unemployed. However, there is also considerable evidence of cause and effect – that work is beneficial, unemployment is harmful, and re-employment promotes health and well-being. There is continued debate about the mechanisms by which these effects occur: through socio-economic status, psychosocial effects, 'stress', altered health behaviour or subsequent employment patterns. It appears likely that all of these mechanisms play a part in different individuals in different contexts for different outcomes. Nevertheless, whatever the exact mechanisms, the outcome is clear: work is generally good and unemployment is generally bad for health and well-being.

Although that general conclusion is clear, there are qualifications in the detail. First, work is only one of many, and often not the most important of the influences on health and well-being. Second, most of this evidence is about group or average effects and there is limited evidence on effect sizes. The impact of work or unemployment on health varies in different individuals in different contexts: effect sizes vary and there may be insufficient evidence to predict them; an important minority of individuals, possibly 5-10%, may show contrary effects. In addition, the findings are based on studies performed over several decades, during which the nature of work may have changed (though it is not clear whether for better or worse). Third, much of the evidence is about short-term effects (~1 year, which is really very short for health impacts) and there is less evidence on longer-term effects over a lifetime perspective. It is not possible to comment further on whether or how long-term might differ from short-term effects. Fourth, the available evidence is mainly about paid employment and unemployment, but many of the findings may be equally applicable to all forms of work and worklessness. It is not possible to say whether paid employment is what matters for health, or if any form of purposeful and meaningful 'work' may be equally good. Finally, there is difficulty distinguishing the health impact of work(lessness) in older workers from the effects of aging, the normal transition to retirement, and the social context in which these occur. It is therefore not possible to draw definite conclusions about whether (early) retirement will be good or bad for an individual's health. Importantly, however, there is no evidence that continued working is generally harmful to the health of older workers, and it may be beneficial. So decisions about retirement age can properly be made on social, economic and other non-health grounds.

The evidence to this point relates to unemployed people who are generally healthy: the previously unconsidered question is whether these conclusions are equally true for sick and disabled people? The evidence on this comes from different sources. Sickness is addressed in the clinical literature, where the primary focus is on clinical management and clinical outcomes and only secondarily about work. Disability is addressed in the disability literature, where the primary focus is on the right to work and full participation in society. Nevertheless, these literatures show a broad consensus across multiple disciplines, disability groups, employers, unions, insurers, and policy makers. When their health condition permits, sick and disabled people should remain in or (re)-enter work as soon as possible because it variously:

- is therapeutic;

- helps to promote recovery and rehabilitation;

- leads to better health outcomes;

- minimises the harmful physical, mental and social effects of long-term sickness absence;

- reduces the risk of long-term incapacity;

- promotes full participation in society, independence and human rights;

- reduces poverty;

- improves quality of life and well-being.

This list combines two sets of evidence: clinical management is based on extensive clinical evidence and experience; disability rights are based on social justice and fairness. There may be little direct scientific evidence that work has a beneficial impact on the health of sick or disabled people, but valid consensus can be established on these other grounds.

The evidence on the main common health problems - mental health, musculoskeletal and cardio-respiratory conditions – supports these findings. In each of the three areas, modern approaches to management emphasise staying active, restoring function, enabling and supporting sick and disabled people to participate in society as fully as possible. These principles apply equally to clinical and occupational management, and lead to a clear goal of work retention and (early) return to work. Each of these conditions provides further evidence to support that consensus. There may again be little direct evidence of the causal link between (return to) work and improved physical and mental health outcomes, but there is a clear association between better clinical and occupational outcomes. Importantly, there is no evidence that work has adverse effects on physical and mental health outcomes, except in very specific circumstances. The strongest arguments then lie in the benefits of work for general and social well-being.

The evidence on social security beneficiaries is somewhat different, and comes from a completely different literature, in which most studies are pragmatic rather than strictly scientific. Nevertheless, and perhaps surprisingly, this search provided some of the clearest and most specific evidence. Re-entering work (rather than simply moving off benefits) is generally associated with improvement in income, socio-economic status, mental and general health, and well-being. This is again partly a health selection effect, but there is also evidence that (re)-entering work improves these outcomes.

Thus, the balance of the evidence is that work is generally good for health and well-being, not only for healthy people, but also for many disabled people, for many people with common health problems, and for many social security beneficiaries. However, that depends on the nature and quality of the job, and the social context. These areas are beyond the scope of this review and are reviewed comprehensively elsewhere (Acheson *et al.* 1998; Ritchie *et al.* 2005; Marmot & Wilkinson 2006), but they are critical to placing the present analysis of work and health in context.

Firstly, work is generally good for your health and well-being, **provided** you have 'a good job'. Good jobs are obviously better than bad jobs, but bad jobs might be either less beneficial or even harmful. It is then important to consider what constitutes a good job. Under UK and European legislation, employers have a statutory duty to conduct suitable risk assessments to identify hazards to health and safety, and to reduce the risks to employees as far as reasonably practicable. But health and safety should be distinguished. As well as controlling risks, it is equally important to make jobs accommodating of common health problems, sickness and disability. A 'healthy working life' goes even further: it is *one that continuously provides working-age people with the opportunity, ability, support and encouragement to work in ways and in an environment which allows them to sustain and improve their health and well-being'* (Scottish Executive 2004). *'Work should be comfortable when we are well and accommodating when we are ill'* (Hadler 1997). The evidence reviewed here suggests that, in terms of promoting health and well-being, the characteristics that distinguish 'good' jobs and 'good' workplaces might include:

- safety

- fair pay

- social gradients in health

- job security

- personal fulfilment and development; investing in human capital

- accommodating, supportive & non-discriminatory

- control/autonomy

- job satisfaction

- good communications

This list is tentative, but clearly goes far beyond physical and mental exposures, demands and risks. Further research is required into the characteristics of a 'good' job, and further consideration is required of the links between good jobs, health and productivity to support the principle that *'good health is good business'.*

There is a particular concern that various physical and psychosocial characteristics of work may be risk factors for some common health problems and hence that (re)-entering work may cause (further) harm to people with these health conditions. However, common health problems have a high prevalence in the normal working age population, causation is multifactorial, effect sizes of work exposures are generally small, and work only accounts for a modest proportion of the impact of these conditions in workers. Psychosocial factors also play an important role in workability and well-being. Physical and mental activity and work can be therapeutic for many common health problems. Thus, provided the 'risks' of work are properly assessed and controlled, provided the demands of work are adjusted where necessary to match individual capacity, and except with very specific conditions and exposures, the beneficial effects of work on physical and mental health and well-being generally outweigh the risks. Supporting this conclusion, the evidence about the beneficial effects of work comes from pragmatic studies of current and past working practice, including its hazards.

Secondly, the relationship between work(lessness) and health must take account of the social context. As a simple example, the impact of work and unemployment varies across age and allowance must be made for the different socio-economic context of school-leavers and of older workers who may be approaching retirement. More fundamentally, there are powerful links between worklessness, poverty, social disadvantage, and exclusion, social inequalities in health, regional deprivation, sickness, and incapacity. There is still a powerful social gradient in health. Socially disadvantaged people are less likely to attain full health and well-being, while chronically sick or disabled people are less likely to fulfil socio-economic roles, leading to poverty. There is also a major geographical dimension around deprived areas, high local unemployment rates, limited job availability, and poverty. People in these areas face multiple personal, health-related, and social disadvantages and barriers to work. It is all very well to say that work is good for your health, but that depends on being able to get a job.

The various findings and issues from this review reinforce the need to develop a more balanced model of the relationship between work and health (Figure 2), which should embody the following principles:

- Safety and health at work should be distinguished;

- There are important interactions between workers and their work, which can modulate any health effects;

- Elements of work can have both beneficial and harmful effects on physical and mental health and well-being;

- Common health problems are usually not a simple 'consequence' of work exposures, but occur in the context of wider interactions between the person (worker) and their (work) environment;

- Common health problems, work and the relationship between them are partly matters of perceptions: the more subjective the problem, the more central the role of psychosocial factors;

- Understanding and addressing common health problems requires a biopsychosocial approach that takes account of the person, their health problem and their work environment.

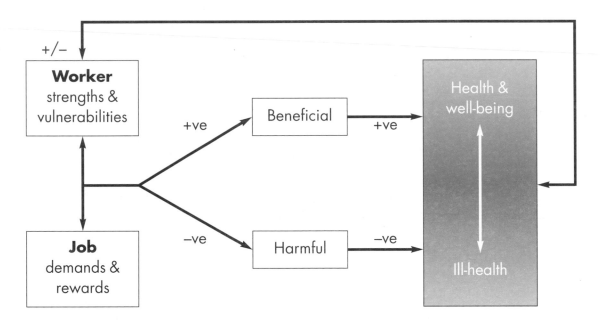

Figure 2. Work and health: interactions can lead to differing consequences

CONCLUSIONS

To answer the specific questions set for this review:

1. *Does the current evidence suggest that work is beneficial for physical and mental health and well being, in general and for common health problems?*

   Yes, work is generally good for the physical and mental health and well-being of healthy people, many disabled people and most people with common health problems. Work can be therapeutic for people with common health problems. Work can reverse the adverse health effects of unemployment.

2. *What is the balance of benefits and risks to health from work and from worklessness?*

In general, provided due care is taken to make jobs as safe and 'good' as reasonably practicable, employment can promote health and well-being, and the benefits outweigh any 'risks' of work and the adverse effects of (long-term) unemployment or sickness absence.

3. *Are there any circumstances (specific people, health conditions or types of work) where work is likely to be detrimental to health and well-being?*

There are some people whose health condition or disability makes it unreasonable to expect them to seek or to be available for work (i.e they fulfil the criteria for entitlement to incapacity benefits) but that does not necessarily imply that work would be detrimental. There are a few people with specific health conditions who should not be exposed to specific occupational hazards (e.g. occupational asthma). However, for healthy people, many disabled people and most people with common health problems, 'good' jobs, if necessary with appropriate accommodations and adjustments, should not be detrimental to health and well-being. The likely benefits outweigh any potential risks.

4. *Are there specific areas where there is a lack of evidence and need for further research?*

Although the broad conclusions of this review are clear, several important issues need further clarification:

- There is limited evidence on effect sizes and a need for further quantitative research: How *much* is work good for health? How *much* does unemployment harm health?

- Most of the evidence is relatively short-term (~1 year) and there is a need for more long-term studies over a lifetime perspective.

- There is a need for further studies of the relative importance, effect sizes, optimum combination and measures of the physical and psychosocial characteristics of jobs that are 'good' for health;

- There is a need for further high quality scientific studies of the impact of work on the health of working-age adults, including the cause and effect relationship, and the relative balance of adverse / beneficial effects of different elements of work;

- There is a need for longitudinal studies to establish and quantify the relative balance of adverse / beneficial effects of (early return to) work vs. continued sickness absence on the physical and mental health of people with common health problems;

- There is a need for longitudinal studies of the relative balance of adverse / beneficial effects of (early) retirement vs. continued working on the physical and mental health of older workers;

- There should be more rigorous research into the whole area of work-related 'stress': further development of basic concepts, definitions, methods of measurement and diagnosis; direct studies of cause & effect relationships and the relative balance of adverse / beneficial effects of (various psychosocial characteristics of) work on mental health.

In summary, despite the diverse nature of the evidence and its limitations in certain areas, this review has built a strong evidence base showing that work is generally good for physical and mental health and well-being. Worklessness is associated with poorer physical and mental health and well-being. Work can be therapeutic and can reverse the adverse health effects of unemployment. That is true for healthy people of working age, for many disabled people, for most people with common health problems and for social security beneficiaries. The provisos are that account must be taken of the social context, the nature and quality of work, and the fact that a minority of people may experience contrary effects. Jobs should be safe and should also be accommodating for sickness and disability. Yet, overall, the beneficial effects of work outweigh the risks of work, and are greater than the harmful effects of long-term unemployment or prolonged sickness absence. Work **is** generally good for health and well-being.

References

AAOS. 2000. *AAOS Position statement. Early return to work programs.* American Academy of Orthopaedic Surgeons, **www.aaos.org/wordhtml/papers/postion/1150htm** (accessed 02 February 2006).

AARP. 2001. *Health and safety issues in an aging workforce.* AARP (Public Policy Institute), Washington, DC.

Aarts LJM, Vurkhauser RV, de Jong PR. 1996. Curing the Dutch disease. In *International Studies of Social Security Vol 1* Avebury, Aldershot.

Abenhaim L, Rossignol M, Valat JP, Nordin M, Avouac B, Blotman F, Charlot J, Dreiser RL, Legrand E, Rozenberg S, Vautravers P. 2000. The role of activity in the therapeutic management of back pain. Report of the International Paris Task Force on back pain. *Spine* 25 (4S): 1S-33S.

ABI. 2003. *Whiplash associated disorder training pack.* Association of British Insurers, London.

ABI/TUC. 2002. *Getting back to work: a rehabilitation discussion paper.* Association of British Insurers, London.

ACC and the National Health Committee. 1997. *New Zealand acute low back pain guide.* Ministry of Health, Wellington, NZ.

Acheson D, Barker D, Chambers J, Graham H, Marmot M, Whitehead M. 1998. *Independent inquiry into Inequalities in Health Report.* The Stationery Office, London.

ACOEM. 2002. *ACOEM consensus opinion statement. The attending physician's role in helping patients return to work after an illness or injury.* American College of Occupational and Environmental Medicine, **www.acoem.org/guidelines/pdf/Return-to-Work-04-02.pdf** (accessed 04 November 2005).

ACOEM. 2005. *Preventing needless work disability by helping people stay employed.* American College of Occupational & Environmental Medicine, **www.acoem.org** (accessed 16 December 2005).

Adisesh A. 2003. Occupational health practice. In *ABC of occupational and environmental medicine* (Ed. Snashall D, Patel D) : 6-11, BMJ Books, London.

Allen M, Barron S, Broudo M, Lubin S, Anton H. 1997. *British Columbia Whiplash Initiative - a comprehensive syllabus.* Physical Medicine Research Foundation, Vancouver, BC.

ARMA. 2004. *Standards of care.* Arthritis and Musculoskeletal Alliance, London **www.arma.uk.net** (accessed 17 February 2006).

Ashworth K, Hartfree Y, Stephenson A. 2001. *Well enough to work?* (DSS Research Report No 145). Corporate Document Services, Leeds.

Asthma UK. 2004. *Asthma at Work - Your Charter.* asthmaUK, **www.asthma.org.uk** (accessed 17 March 2006).

Backman CL. 2004. Employment and work disability in rheumatoid arthritis. *Curr Opin Rheumatol* 16: 148-152.

Bacon J. 2002. Moving between sickness and unemployment. *Labour Market Trends* 110: 195-205.

Baltes RB, Briggs TE, Huff JW, Wright JA, Neuman GA. 1999. Flexible and compressed workweek schedules: a meta-analysis of their effects on work-related criteria. *Journal of Applied Psychology* 84: 496-513.

Banks MH. 1995. Psychological effects of prolonged unemployment: relevance to models of work re-entry following injury. *Journal of Occupational Rehabilitation* 5: 37-53.

Banks MH, Jackson PR. 1982. Unemployment and risk of minor psychiatric disorder in young people: cross-sectional and longitudinal evidence. *Psychological Medicine* 12: 789-798.

Barham C. 2002. Economic inactivity and the labour market. *Labour Market Trends* 69-77.

Baronet A-M, Gerber GJ. 1998. Psychiatric rehabilitation: efficacy of four models. *Clinical Psychology Review* 18: 189-228.

Bartley M. 1994. Unemployment and ill health: understanding the relationship. *Journal of Epidemiology and Community Health* 48: 333-337.

Bartley M, Sacker A, Schoon I, Kelly MP, Carmona C. 2005. *Work, non-work, job satisfaction and psychological health: evidence review.* Health Development Agency, **www.publichealth.nice.org.uk** (accessed 18 January 2006).

Barton R. 1999. Psychosocial rehabilitation services in community support systems: a review of outcomes and policy recommendations. *Psychiatric Services* 50: 525-534.

Benavides FG, Benach J, Diez-Roux AV, Roman C. 2000. How do types of employment relate to health indicators? Findings from the Second European Survey on Working Conditions. *J Epidemiol Community Health* 54: 494-501.

Benjamin K, Wilson S. 2005. *Facts and misconceptions about age, health status and employability.* Health and Safety Laboratory, Buxton: **http://www.agepositive.gov.uk/agepartnershipgroup/research/agehealthemployability.pdf** (accessed 16 December 2005).

Berthoud R. 2003. *Multiple disadvantage in employment.* Joseph Rowntree Foundation, London.

BICMA. 2000. *Code of best practice on rehabilitation, early intervention and medical treatment in personal injury claims: a practitioner's guide to rehabilitation.* Bodily Injury Claims Management Association, London.

Bjarnason T, Sigurdardottir TJ. 2003. Psychological distress during unemployment and beyond: social support and material deprivation among youth in six northern European countries. *Social Science & Medicine* 56: 973-985.

Björklund A, Eriksson T. 1998. Unemployment and mental health: evience from research in the Nordic countries. *Scand J Soc Welfare* 7: 219-235.

Bloch FS, Prins R. 2001. *Who return to work & why? A six-country study on work incapacity & reintegration.* Transaction Publishers, New Brunswick.

Bond FW, Flaxman PE, Loivette S. 2006. *A business case for the Management Standards for stress. (HSE RR 431).* HSE Books, London.

Bond GR. 2004. Supported employment: evidence for an evidence-based practice. *Psychiatric Rehabilitation Journal* 27: 345-359.

Bond GR, Resnick SG, Drake RE, Xie H, McHugo GJ, Bebout RR. 2001. Does competitive employment improve nonvocational outcomes for people with severe mental illness? *Journal of Consulting and Clinical Psychology* 69: 489-501.

Bound J. 1989. The health and earnings of rejected disability insurance applicants. *American Economic Review* 79: 482-503.

Bowen J, Lunt J, Lee R. 2005. *Report on procedings of HSE's Health Models Review workshop - Manchester 20 Spetember 2005 (MU/06/03).* Health & Safety Laboratory/ Health & Safety Executive, London.

Bowling J, Coleman N, Wapshott J, Carpenter H. 2004. *Destination of benefit leavers (In-House Report 132).* Department for Work and Pensions, London.

Branthwaite A, Garcia S. 1985. Depression in the young unemployed and those on Youth Opportunities Schemes. *British Journal of Medical Psychology* 58: 67-74.

Brenner MH. 2002. *Employment and public health. Final report to the European Commission Directorate General Employment, Industrial Relations and Social Affairs. Volume I-III.* European Commission, Brussels
http://europa.eu.int/comm/employment_social/news/2002/jul/empl_health_en.html (accessed 18 January 2006).

Brenner MH, Mooney A. 1983. Unemployment and health in the context of economic change. *Soc Sci Med* 17: 1125-1138.

Brezinka V, Kittel F. 1995. Psychosocial factors of coronary heart disease in women: a review. *Soc Sci Med* 42: 1351-1365.

Briner RB. 2000. Relationships between work environments, psychological environments and psychological well-being. *Occupational Medicine* 50: 299-303.

Briner RB, Reynolds S. 1999. The costs, benefits, and limitations of organizational level stress interventions. *Journal of Organizational Behavior* 20: 647-664.

BSRM. 2000. *Vocational rehabilitation. The way forward*. British Society of Rehabilitation Medicine, London.

Buckle P, Devereux J. 1999. *Work-related neck and upper limb musculoskeletal disorders*. European Agency for Safety and Health at Work, Luxembourg.

Burchell B. 1994. The effects of labour market position, job insecurity, and unemployment on psychological health. In *Social change and the experience of unemployment* (Ed. Gallie D, Marsh C, Vogler C) : 188-212, Oxford University Press, Oxford.

Burton AK. 1997. Back injury and work loss. Biomechanical and psychosocial influences. *Spine* 22: 2575-2580.

Burton AK, Balagué F, Eriksen HR, Henrotin Y, Lahad A, Leclerc A, Mûller G, van der Beek AJ, on behalf of the COST B13 Working Group on Guidelines for Prevention in Low Back Pain. 2004. *European guidelines for prevention in low back pain*. EC Cost Action B13, **www.backpaineurope.org**

Burton W, Morrison A, Maclean R, Ruderman E. 2006. Systematic review of studies of productivity loss due to rheumatoid arthritis. *Occupational Medicine* 56: 18-27.

Cairns R, Hotopf M. 2005. A systematic review describing the prognosis of chronic fatigue syndrome. *Occupational Medicine* 55: 20-31.

Caplan RD, Vinokur AD, Price RH, van Ryn M. 1989. Job seeking, reemployment, and mental health: a randomized field experiment in coping with job loss. *Journal of Applied Psychology* 74: 759-769.

Carter JT, Birrell LN. 2000. *Occupational health guidelines for the management of low back pain at work - principal recommendations*. Faculty of Occupational Medicine, London.

Catalano R. 1991. The health effects of economic insecurity. *American Journal of Public Health* 81: 1148-1152.

CBI. 2000. *Their health in your hands. Focus on occupational health partnerships*. Confederation of British Industry, London.

Chirikos TN. 1993. The relationship between health and labor market status. *Annu Rev Public Health* 14: 293-312.

Claussen B. 1999. Alcohol disorders and re-employment in a 5-year follow-up of long-term unemployed. *Addiction* 94: 133-138.

Claussen B, Bjørndal A, Hjort PF. 1993. Health and re-employment in a two year follow up of long term unemployed. *Journal of Epidemiology and Community Health* 47: 14-18.

Coats D. 2005. *An agenda for work: The Work Foundation's challenge to policy makers.* The Work Foundation, London **www.theworkfoundation.com/research/agenda.jsp** (accessed 05 April 2006).

Coats D, Max C. 2005. *Healthy work: productive workplaces. Why the UK needs more "good jobs".* The Work Foundation, London **www.theworkfoundation.com/research/healthy_work.jsp** (accessed 19 January 2006).

Coggon D. 1994. The epidemiology of occupational disease. In *Hunter's diseases of occupations* (Ed. Raffle PAB, Adams PH, Baxter PJ, Lee WR) : 49-56, Edward Arnold, London.

Cohen S. 1999. Social status and susceptibility to respiratory infections. *Annals New York Academy of Sciences* 896: 246-253.

Cohn RM. 1978. The effect of employment status change on self-attitudes. *Social Psychology* 41: 81-93.

Coleman N, Kennedy L. 2005. *Destination of benefit leavers 2004 (DWP Research Report No 244).* Corporate Document Services, Leeds.

Corden A, Thornton P. 2002. *Employment programmes for disabled people: lessons from research evaluations (In-house report 90).* Her Majesty's Stationery Office, London.

COST B13 working group. 2004. *Low back pain: guidelines for its management.* European Commission, **www.backpaineurope.org** (accessed 06 February 2006).

Coulter A, Entwistle V, Gilbert D. 1998. *Informing patients. An assessment of the quality of patient information materials.* Kings Fund, London.

Council of Europe. 1992. *Recommendation No. R (92) 6: On a coherent policy for people with disabilities.* Council of Europe, **www.coe.int** (accessed 18 January 2006).

Cox T. 1993. *Stress research and stress management: putting theory to work (HSE CRR 061).* HSE Books, London.

Cox T, Griffiths A, Barlowe C, Randall R, Thomson L, Rial-Gonzalez E. 2000a. *Organisational interventions for work stress. A risk management approach (HSE CRR 286).* HSE Books, London.

Cox T, Griffiths A, Rial-González E. 2000b. *Research on work-related stress*. European Agency for Safety and Health at Work, Luxembourg.

Cox T, Leka S, Ivanov I, Kortum E. 2004. Work, employment and mental health in Europe. *Work & Stress* 18: 179-185.

Cox T, Tisserand M, Tarls T. 2005. The conceptualisation and measurement of burnout: questions and directions. Editorial. *Work & Stress* 19: 187-191.

Crowley JE. 1986. Longitudinal effects of retirement on men's well-being and health. *Journal of Business and Psychology* 1: 95-113.

Crowther R, Marshall M, Bond G, Huxley P. 2001a. *Vocational rehabilitation for people with severe mental illness. The Cochrane Database of Systematic Reviews*. John Wiley & Sons Ltd, Chicester.

Crowther RE, Marshall M, Bond GR, Huxley P. 2001b. Helping people with severe mental illness to obtain work: systematic review. *BMJ* 322: 204-208.

D'Souza JC, Franzblau A, Werner RA. 2005. Review of epidemiologic studies on occupational factors and lower extremity musculoskeletal and vascular disorders and symptoms. *Journal of Occupational Rehabilitation* 15: 129-165.

Dafoe WA, Cupper L. 1995. Vocational considerations and return to work. *Physical Medicine and Rehabilitation Clinics of North America* 6: 191-204.

Daniel WW. 1983. How the unemployed fare after they find new jobs. *Policy Studies* 3: 246-260.

Danna K, Griffin RW. 1999. Health and well-being in the workplace: a review and synthesis of the literature. *Journal of Management* 25: 357-384.

Davis KG, Heaney CA. 2000. The relationship between psychosocial work characteristcs and low back pain: underlying methodological issues. *Clin Biomech* 15: 389-406.

De Beek RO, Hermans V. 2000. *Research on work-related low back disorders*. European Agency for Safety and health, Luxembourg.

de Buck PDM, Schoones JW, Allaire SH, Vliet Vlieland TPM. 2002. Vocational rehabilitation in patients with chronic rheumatic diseases: a systematic literature review. *Seminars in Arthritis and Rheumatism* 32: 196-203.

de Croon EM, Sluiter JK, Nijssen TF, Dijkmans BAC, Lankhorst GJ, Frings-Dresen MHW. 2004. Predictive factors of work disability in rheumatoid arthritis: a systematic literature review. *Ann Rheum Dis* 63: 1362-1367.

De Dreu CKW, Weingart LR. 2003. Task versus relationship conflict, team performance, and team member satisfaction: a meta-analysis. *Journal of Applied Psychology* 88: 741-749.

de Gaudemaris R. 2000. Clinical issues: return to work and public safety. *Occupational Medicine: State of the Art Reviews* 15: 223-230.

de Lange AH, Taris TW, Kompier MAJ, Houtman ILD. 2003. "The *very* best of the millennium": Longitudinal research and the demand-control-(support) model. *Journal of Occupational Health Psychology* 8: 282-305.

Deacon A. 1997. *From welfare to work: lessons from America*. Institute of Economic Affairs, Health & Welfare Unit, London.

Deacon S. 2005. *Health and performance: a guide to employee health and productivity in the workplace*. AXA PPP Healthcare, Tunbridge Wells.

Dean H. 2003. Re-conceptualising welfare-to-work for people with multiple problems and needs. *J Social Policy* 32: 441-459.

Department of Health. 2000. *The NHS Plan*. The Stationery Office, London **www.nhsia.nhs.uk/nhsplan** (accessed 11 March 2006).

Detmer DE, MacLeod SA, Wait S, Taulor M, Ridgwell J. 2003. *The informed patient: Study Report*. Cambridge University Health, Cambridge.

DH. 2000. *National Service Framework for coronary heart disease: modern standards and service models*. Department of Health, London.

DH. 2004. *Choosing health: making healthier choices easier: Public Health White Paper. Cm 6374*. Department of Health, London.

Diener E. 2000. Subjective well-being. The science of happiness and a proposal for a national index. *American Psychologist* 55: 34-43.

Disability Rights Commission. 2004. *Strategic plan 2004/5 to 2006/7*. Disability Rights Commission, London.

Dodu N. 2005. Is employment good for well-being? a literature review. *Journal of Occupational Psychology, Employment and Disability* 7: 17-33.

Donovan A, Oddy M, Pardoe R, Ades A. 1986. Employment status and psychological well-being: a longitudinal study of 16-year-old school leavers. *J Child Psychol Psychiat* 27: 65-76.

Dooley D. 2003. Unemployment, underemployment, and mental health: conceptualizing employment status as a continuum. *American Journal of Community Psychology* 32: 9-20.

Dooley D, Prause J. 1995. Effect of unemployment on school leavers' self-esteem. *Journal of Occupational and Organizational Psycholoy* 68: 177-192.

Dorsett R, Finlayson L, Ford R, Marsh A, White M, Zarb G. 1998. *Leaving Incapacity Benefit. Department of Social Security Research Report No 86*. The Stationery Office, London.

DRC. 2004a. *Conditions for conditionality*. Disability Rights Commission, London **www.drc-gb.org** (accessed 11 April 2006).

DRC. 2004b. *Disability briefing*. Disability Rights Commission, London **www.drc-gb.org** (accessed 11 April 2006).

DRC. 2004c. *Disability Rights Commission Strategic Plan 2004-2007*. Disability Rights Commission, London **www.drc-gb.org** (accessed 11 April 2006).

Dunn AL, Trivedi MH, O'Neal HA. 2001. Physical activity dose-response effects on outcomes of depression and anxiety. *Medicine & Science in Sports & Exercise* 33 (Suppl): S587-S597.

DWP. 2003. *Patients, their employment and their health -how to help your patients stay in work*. Department for Work and Pensions, London. **http://www.dwp.gov.uk/medical/hottopics/dwp-desk-aid-time-line-2003-4.pdf** (accessed 3 January 2006).

DWP. 2004a. *Building capacity for work: a UK framework for vocational rehabilitation*. Department for Work and Pensions, London.

DWP. 2004b. *IB204: Medical evidence for statutory sick pay, statutory maternity pay and social security incapacity benefit purposes - a guide for registered medical practitioners*. Department for Work and Pensions, London. **http://www.dwp.gov.uk/medical/medicalib204/ib204-june04/ib204.pdf** (accessed 3 January 2006).

DWP. 2006. *A new deal for welfare: Empowering people to work*. Department for Work & Pensions, London.

Dwyer P. 2000. *Welfare rights and responsibilities: contesting social citizenship*. Policy Press, Bristol.

Edwards JR, Cooper CL. 1988. The impacts of positive psychological states on physical health: a review and theoretical framework. *Soc Sci Med* 27: 1447-1459.

EFILWC. 2004. *Employment and disability: back to work strategies - summary*. European Foundation for the Improvement of Living and Working Conditions, Dublin. **www.eurofound.eu.int/publications/files/EF041152EN.pdf** (accessed 9 December 2005), **www.eurofound.eu.int/publications/files/EF0499EN.pdf** (accessed 9 December 2005).

Eisenberg P, Lazarsfeld F. 1938. The psychological effects of unemployment. *Psychol Bull* 35: 358-390.

Ekerdt DJ, Bosse R, LoCastro JS. 1983. Claims that retirement improves health. *Journal of Gerontology* 38: 231-236.

Erens B, Ghate D. 1993. *Invalidity benefit: a longitudinal survey of new recipients (DSS Research Report No 20)*. Her Majesty's Stationery Office, London.

Eriksen HR, Ursin H. 1999. Subjective health complaints: is coping more important than control? *Work & Stress* 13: 238-252.

Eriksen HR, Ursin H. 2002. Sensitization and subjective health complaints. *Scand J Psychology* 43: 189-196.

Eriksen HR, Ursin H. 2004. Subjective health complaints, sensitization, and sustained cognitive activation (stress). *Journal of Psychosomatic Research* 56: 445-448.

Ezzy D. 1993. Unemployment and mental health: a critical review. *Soc Sci Med* 37: 41-52.

Faragher EB, Cass M, Cooper CL. 2005. The relationship between job satisfaction and health: a meta-analysis. *Occup Environ Med* 62: 105-112.

Feather NT, O'Brien GE. 1986. A longitudinal analysis of the effects of different patterns of employment and unemployment on school-leavers. *British Journal of Psychology* 77: 459-479.

Felce D, Perry J. 1995. Quality of life: its definition and measurement. *Research in Developmental Disibilities* 16: 51-74.

Felson DT. 1994. Do occupation-related physical factors contribute to arthritis? *Baillière's Clinical Rheumatology* 8: 63-77.

Ferguson SS, Marras WS. 1997. A literature review of low back disorder surveillance measures and risk factors. *Clin Biomech* 12: 211-226.

Ferrie JE. 1999. Health consequences of job insecurity. In *Labour market changes and job insecurity: a challenge for social welfare and health promotion (WHO Regional Publications, European Series, No. 81)* (Ed. Ferrie JE, Marmot MG, Griffiths J, Ziglio E) WHO, Copenhagen.

Ferrie JE. 2001. Is job insecurity harmful to health? *J R Soc Med* 94: 71-76.

Ferrie JE, Martikainen P, Shipley MJ, Marmot MG, Stansfeld SA, Smith GD. 2001. Employment status and health after privatisation in white collar civil servants: prospective cohort study. *BMJ* 322: 647-651.

Ferrie JE, Shipley MJ, Stansfeld SA, Marmot MG. 2002. Effects of chronic job insecurity and change in job security on self reported health, minor psychiatric morbidity, physiological measures, and health related behaviours in British civil servants: the Whitehall II study. *J Epidemiol Community Health* 56: 450-454.

Field F. 1998. *Reflections on welfare reform*. The Social Market Foundation, London.

Fifield J, Reisine ST, Grady K. 1991. Work disability and the experience of pain and depression in rheumatoid arthritis. *Soc Sci Med* 33: 579-585.

FOM. 2005. *The health and work handbook*. Faculty of Occupational Medicine, Royal College of General Practitioners, Society of Occupational Medicine, London. **www.facoccmed.ac.uk/library/docs/h&w.pdf** (accessed 9 December 2005).

Ford FM, Ford J, Dowrick C. 2000. Welfare to work: the role of general practice. *Brit J Gen Practice* 50: 497-500.

Fordyce WE. 1995. *Back pain in the workplace: management of disability in nonspecific conditions - a report of the Task Force on Pain in the Workplace of the International Association for the Study of Pain*. IASP Press, Seattle.

FPH/FOM. 2006. *Creating healthy workplace. A guide for occupational safety and health professionals and employers*. Faculty of Public Health and Faculty of Occupational Medicine, **www.fph.org.uk** and **www.facoccmed.ac.uk** (accessed 20 March 2006).

Franche RL, Cullen K, Clarke J, Irvin E, Sinclair S, Frank J, The Institute for Work & Health (IWH) Workplace-Based RTW Intervention Literature Review Research Team. 2005. Workplace-based return-to-work interventions: a systematic review of the quantitative literature. *J Occup Rehabil* 15: 607-631.

Frank J, Sinclair S, Hogg-Johnson S, Shannon H, Bombardier C, Beaton D, Cole D. 1998. Preventing disability from work-related low-back pain. New evidence gives new hope - if we can just get all the players onside. *Canadian Medical Association Journal* 158: 1625-1631.

Frank JW, Brooker AS, DeMaio SE, Kerr MS, Maetzel A, Shannon HS, Sullivan TJ, Norman RW, Wells RP. 1996. Disability resulting from occupational low back pain: Part II: what do we know about secondary prevention? a review of the scientific evidence on prevention after disability begins. *Spine* 21: 2918-2929.

Franklin BA, Bonzheim K, Gordon S, Timmis GC. 1998. Rehabilitation of cardiac patients in the twenty-first century: Changing paradigms and perceptions. *Journal of Sports Sciences* 16: S57-S70.

Frese M. 1987. Alleviating depression in the unemployed: adequate financial support, hope and early retirement. *Soc Sci Med* 25: 213-215.

Frese M, Mohr G. 1987. Prolonged unemployment and depression in older workers: a longitudinal study of intervening variables. *Soc Sci Med* 25: 173-178.

Fryers T, Melzer D, Jenkins R. 2003. Social inequalities and the common mental disorders. A systematic review of the evidence. *Soc Psychiatr Epidemiol* 38: 229-237.

Gallo WT, Bradley EH, Falba TA, Dubin JA, Cramer LD, Bogardus Jr ST, Kasl SV. 2004. Involuntary job loss as a risk factor for subsequent myocardial infarction and stroke: findings from the health and retirement survey. *American Journal of Industrial Medicine* 45: 408-416.

Gallo WT, Bradley EH, Siegel M, Kasl SV. 2000. Health effects of involuntary job loss among older workers: findings from the health and retirement survey. *Journal of Gerontology* 55B: S131-S140.

Gallo WT, Bradley EH, Siegel M, Kasl SV. 2001. The impact of involuntary job loss on subsequent alcohol consumption by older workers: findings from the health and retirement survey. *Journal of Gerontology* 56B: S3-S9.

Garman A, Redmond G, Lonsdale S. 1992. *Incomes in and out of work. A cohort study of newly unemployed men and women (DSS Research Report No 7)*. Her Majesty's Stationery Office, London.

Gibson PG, Powell H, Coughlan J, Wilson AJ, Abramson M, Haywood P, Bauman A, Hensley MJ, Walters EH. 2003. Self-management education and regular practitioner review for adults with asthma (Cochrane Review). In *The Cochrane Library, Issue 1* Update Software, Oxford.

Glozier N. 2002. Mental ill health and fitness for work. *Occup Environ Med* 59: 714-720.

Goldstone C, Douglas L. 2003. *Pathways to Work from Incapacity Benefits: A pre-pilot exploration of staff and customer attitudes. Report prepared for DWP*. Department for Work and Pensions, London.

Graetz B. 1993. Health consequences of employment and unemployment: longitudinal evidence for young men and women. *Soc Sci Med* 36: 715-724.

Green F. 2003. *The rise and decline of job insecurity. Department of Economics Discussion Paper 2003*. University of Kent, Canterury **www.kent.ac.uk/economics/papers** (accessed 18 January 2006).

Grewal I, McManus S, Arthur S, Reith L. 2004. *Making transition: addressing barriers in services for disabled people (DWP Research Report No 204)*. Corporate Document Services, Leeds.

Haafkens J, Moerman C, Schuring M, van Dijk F. 2006. Searching bibliographic databases for literature on chronic disease and work participation. *Occup Med* 56: 39-45.

Hadler NM. 1997. Back pain in the workplace. What you lift or how matters far less than whether you lift or when. *Spine* 22: 935-940.

Hakim C. 1982. The social consequences of high unemployment. *Journal of Social Policy* 11: 433-467.

Haldorsen EMH, Brages S, Johannessen TS. 1996. Musculoskeletal pain: concepts of disease, illness and sick certification in health professionals in Norway. *Scand J Rheumatol* 25: 224-232.

Halvorsen K. 1998. Impact of re-employment on psychological distress among long-term unemployed. *Acta Sociologica* 41: 227-242.

Hamilton VH, Merrigan P, Dufresne É. 1997. Down and out: estimating the relationship between mental health and unemployment. *Health Economics* 6: 397-406.

Hamilton VL, Hoffman WS, Broman CL, Rauma D. 1993. Unemployment, distress, and coping: a panel study of autoworkers. *Journal of Personality and Social Psychology* 65: 234-247.

Hammarström A. 1994a. Health consequences of youth unemployment. *Public Health* 108: 403-412.

Hammarström A. 1994b. Health consequences of youth unemployment - review from a gender perspective. *Soc Sci Med* 38: 699-709.

Hammarström A, Janlert U. 2002. Early unemployment can contribute to adult health problems: results from a longitudinal study of school leavers. *J Epidemiol Community Health* 56: 624-630.

Hansson RO, DeKoekkoek PD, Neece WM, Patterson DW. 1997. Successful aging at work: annual review, 1992-1996: the older worker and transitions to retirement. *Journal of Vocational Behavior* 51: 202-233.

Haralson RHI. 2005. Working with common lower extremity problems. In *A Physician's Guide to Return to Work* (Ed. Talmage JB, Melhorn JM) : 215-231, American Medical Association, Chicago.

Harnois G, Gabriel P. 2000. *Mental health and work: impact, issues and good practices.* World Health Organization, Geneva (WHO/MSD/MPS/00.2).

Harrington JM. 1994a. Shift work and health - A critical review of the literature on working hours. *Ann Acad Med Singapore* 23: 699-705.

Harrington JM. 1994b. Working long hours and health. *BMJ* 308: 1581-1582.

Haynes SG, McMichael AJ, Tyroler HA. 1978. Survival after early and normal retirement. *Journal of Gerontology* 33: 269-278.

HDA. 2004. *The evidence about work and health.* Health Development Agency, **www.publichealth.nice.org.uk** (accessed 18 January 2006).

Hedges A, Sykes W. 2001. *Moving between sickness and work (DWP Research Report No 151).* Corporate Document Services, Leeds.

Helliwell PS, Taylor WJ. 2004. Repetitive strain injury. *Postgrad Med Journal* 80: 438-443.

Hemingway H, Marmot M. 1999. Psychosocial factors in the aetiology and prognosis of coronary heart disease: systematic review of prospective cohort studies. *BMJ* 318: 1460-1467.

Hemp P. 2004. Presenteeism: at work but out of it. *Harvard Business Review* 82: 10-49.

Henriksson CM, Liedberg GM, Gerdle B. 2005. Women with fibromyalgia: work and rehabilitation. *Disability and Rehabilitation* 27: 685-695.

HM Government. 2005. *Health, work and well-being - caring for our future.* Department for Work and Pensions, London **www.dwp.gov.uk/publications/dwp/2005/health_and_wellbeing.pdf** (accessed 18 January 2006).

Hoogendoorn WE, van Poppel MNM, Bongors PM, Koes BW, Bouter LM. 1999. Physical load during work and leisure time as risk factors for back pain. *Scand J Work Environ Health* 25: 387-403.

Horgan J, Bethell H, Carson P, Davidson C, Julian D, Mayou RA, Nagle R. 1992. Working party report on cardiac rehabilitation. *British Heart Journal* 67: 412-418.

Howard M. 2004. *Equal citizenship and Incapacity Benefit reform. Paper to IPPR seminar on the future of incapacity benefits. London 28 October 2004.* Disability Rights Commission, London **www.drc-gb.org** (accessed 11 April 2006).

HSC. 2002. *The health and safety system in Great Britain.* Health and Safety Commission, London. **www.hse.gov.uk/aboutus/hsc** (accessed 01 February 2006).

HSC. 2004. *A strategy for workplace health and safety in Greaat Britain to 2010 and beyond.* Health and Safety Commission, London. **www.hse.gov.uk/aboutus/hsc/strategy.htm** (accessed 01 February 2006).

HSE. 2004a. *Management standards for work-related stress.* Health & Safety Executive, London **http://www.hse.gov.uk/stress/standards/** (accessed 13 December 2005).

HSE. 2004b. *Managing sickness absence and return to work - an employers' and managers' guide.* Health and Safety Executive [HSE Books], London.

HSE. 2005. *Working to prevent sickness absence becoming job loss: practical advice for safety and other trade union representatives.* Health and Safety Executive, London.

HSE. 2006. *Asthma.* Health & Safety Executive, **www.hse.gov.uk/asthma** (accessed 17 March 2006).

HSE/HSL. 2005. *HSE review of the risk prevention approach to occupational health. Applying health models to 21st century occupational health needs - Information pack; Workshop 20 September 2005.* Health & Safety Executive/Health & Safety Laboratory, London.

Hyman MH. 2005. Working with common cardiopulmonary problems. In *A Physician's Guide to Return to Work* (Ed. Talmage JB, Melhorn JM) : 233-266, American Medical Association, Chicago.

Ihlebæk C, Eriksen HR. 2003. Occupational and social variation in subjective health complaints. *Occupational Medicine* 53: 270-278.

IIAC. 2004. *Position Paper No.13 Stress at work.* Industrial Injuries Advisory Council, London **www.iiac.org.uk/papers.shtml** (accessed 18 January 2006).

IIAC. 2006. *Prescribed diseases.* Industrial Injuries Advisory Council, **www.iiac.org.uk/about.shtml** (accessed 11 February 2006).

Ilmarinen JE. 2001. Aging workers. *Occupational and Environmental Medicine* 58: 546-552 doi:10.1136/oem.58.8.546.

ILO. 2002. *ILO code of practice. Managing disability in the workplace.* International Labour Office, Geneva.

IoD. 2002. *Health and wellbeing in the workplace: managing health, safety and wellbeing at work to boost business performance.* Institute of Directors, London.

IUA. 2003. *The rehabilitation code and guide: 4. Report of the rehabilitation working party.* International Underwriting Association, London **www.iua.co.uk** (accessed 06 January 2006).

Iversen L, Sabroe S. 1988. Psychological well-being among unemployed and employed people after a company closedown: a longitudinal study. *Journal of Social Issues* 44: 141-152.

Jackson PR, Stafford EM, Banks MH, Warr PB. 1983. Unemployment and psychological distress in young people: the moderating role of employment commitment. *Journal of Applied Psychology* 68: 525-535.

Jahoda M. 1982. *Employment and unemployment.* Cambridge University Press, Cambridge.

James P, Cunningham I, Dibben P. 2002. Absence management and the issues of job retention and return to work. *Human Resource Management Journal* 12: 82-94.

James P, Cunningham I, Dibben P. 2003. *Job retention and vocational rehabilitation: the development and evaluation of a conceptual framework (HSE RR 106).* HSE Books, London.

Janlert U. 1997. Unemployment as a disease and diseases of the unemployed. *Scan J Work Environ Health* 23: 79-83.

Jin RL, Shah CP, Svoboda TJ. 1995. The impact of unemployment on health: a review of the evidence. *Can Med Assoc J* 153: 529-540.

Juvonen-Posti P, Kallanranta T, Eksymä S-L, Piirainen K, Keinänen-Kiukaanniemi S. 2002. Into work, through tailored paths: a two-year follow-up of the return-to-work rehabilitation and re-employment project. *International Journal of Rehabilitation Research* 25: 313-330.

Kaplan GA, Keil JE. 1993. Socioeconomic factors and cardiovascular disease: a review of the literature. *Circulation* 88: 1973-1998.

Kazimirski A, Adelman L, Arch J, Keenan L, Legge K, Shaw A, Stafford B, Taylor R, Tipping S. 2005. *New deal for disabled people evaluation: registrants' survey - merged cohorts (Cohorts one and two, Waves one and two) (DWP Research Report No 260).* Corporate Document Services, Leeds.

Kazimirski JC. 1997. CMA Policy Summary: The physician's role in helping patients return to work after an illness or injury. *Can Med Assoc J* 156: 680A-680C.

Kendall N. 2003. *Evidence Review: Raising the awareness of key frontline health professionals about the importance of work, job retention, and rehabilitation for their patients (DWP internal document).* Department for Work and Pensions, London.

Kessler RC, Blake Turner J, House JS. 1989. Unemployment, reemployment, and emotional functioning in a community sample. *American Sociological Review* 54: 648-657.

Kilbom Å. 1999. Evidence-based programs for the prevention of early exit from work. *Experimental Aging Research* 25: 291-299.

King D, Wickam-Jones M. 1999. From Clinton to Blair: the Democratic (Party) origins of welfare to work. *Political Quarterly* 70: 62-74.

Klumb PL, Lampert T. 2004. Women, work, and well-being 1950-2000: a review and methodological critique. *Social Science & Medicine* 58: 1007-1024.

Koes BW, van Tulder MW, Ostelo R, Burton AK, Waddell G. 2001. Clinical guidelines for the management of low back pain in primary care: an international comparison. *Spine* 26: 2504-2513.

Kornfeld R, Rupp K. 2000. The net effects of the Project NetWork return-to-work case management experiment on participant earnings, benefit receipt, and other outcomes. *Soc Secur Bull* 63: 12-33.

Kovoor P, Lee AKY, Carrozzi F, Wiseman V, Byth K, Zechin R, Dickson C, King M, Hall J, Ross DL, Uther JB, Dennis AR. 2006. Return to full normal activities including work at two weeks after acute myocardial infarction. *Am J Cardiol* 97: 952-958.

Lacasse Y, Brosseau L, Milne S, Martin S, Wong E, Guyatt GH, Goldstein RS, White J. 2003. Pulmonary rehabilitation for chronic obstructive pulmonary disease (Cochrane Review). In *The Cochrane Library, Issue 4* John Wiley & Sons, Ltd, Chichester.

Lahelma E. 1992. Unemployment and mental well-being: elaboration of the relationship. *International Journal of Health Services* 22: 261-274.

Lakey J. 2001. *Youth unemployment, labour market programmes and health.* Policy Studies Institute, London.

Layard R. 2004. *CEP Occasional Paper No 19 April 2004. Good jobs and bad jobs.* Centre for Economic Performance, London.

Layton C. 1986a. Change-score analyses on the GHQ and derived subscales for male school-leavers with subsequent differing work status. *Person Individ Diff* 7: 419-422.

Layton C. 1986b. Employment, unemployment, and response to the general health questionnaire. *Psychological Reports* 58: 807-810.

Lazarus RS, Folkman S. 1984. *Stress, appraisal, and coping.* Springer Publishing Company, New York.

Lee RT, Ashforth BE. 1996. A meta-analytic examination of the correlates of the three dimensions of job burnout. *J Appl Psychol* 81: 123-133.

Leech C. 2004. *Preventing chronic disability from low back pain - Renaissance Project.* The Stationery Office, Dublin (Government Publications Office).

Liira J, Leino-Arjas P. 1999. Predictors and consequences of unemployment in construction and forest work during a 5-year follow-up. *Scand J Work Environ Health* 25: 42-49.

Locke EA. 1969. What is job satisfaction? *Organizational Behaviour and Human Performance* 4: 309-336.

Loisel P, Buchbinder R, Hazard R, Keller R, Scheel I, van Tulder M, Webster B. 2005. Prevention of work disability due to musculoskeletal disorders: the challenge of implementing evidence. *Journal of Occupational Rehabilitation* 15: 507-521.

Lötters F, Hogg-Johnson S, Burdorf A. 2005. Health status, its perceptions, and effect on return to work and recurrent sick leave. *Spine* 30: 1086-1092.

Lusk SL. 1995. Returning to work following myocardial infarction. *AAOHN* 43: 155-158.

Lynge E. 1997. Unemployment and cancer: a literature review. *IARC Sci Publ* (138): 343-351.

Mackay CJ, Cousins R, Kelly PJ, Lee S, McCaig RH. 2004. 'Management Standards' and work-related stress in the UK: policy background and science. *Work & Stress* 18: 91-112.

Malo J-L. 2005. Future advances in work-related asthma and the impact on occupational health. *Occupational Medicine* 55: 606-611.

Marmot M. 2004. *Status syndrome*. Bloomsbury, London.

Marmot M, Wilkinson RG. 2006. *Social determinants of health (2nd edition)*. Oxford University Press, Oxford.

Marwaha S, Johnson S. 2004. Schizophrenia and employment. A review. *Soc Psychiatry Psychiatr Epidemiol* 39: 337-349.

Mathers CD, Schofield DJ. 1998. The health consequences of unemployment: the evidence. *Med J Aust* 168: 178-182.

McClune T, Burton AK, Waddell G. 2002. Whiplash associated disorders: a review of the literature to guide patient information and advice. *Emergency Medicine Journal* 19: 499-506.

McCluskey S, Burton AK, Main CJ. 2006. The implementation of occupational health guidelines principles for reducing sickness absence due to musculoskeletal disorders. *Occup Med*.

McLean C, Carmona C, Francis S, Wohlgemuth C, Mulvihill C. 2005. *Worklessness and health - what do we know about the casual relationship? Evidence review. 1st edition*. Health Development Agency, London.

Mead LM. 1997. *The new paternalism: supervisory approaches to poverty*. Brookings Institute Press, Washington DC.

Mean Patterson LJ. 1997. Long-term unemployment amongst adolescents: a longitudinal study. *Journal of Adolescence* 20: 261-280.

Mein G, Martikainen P, Hemingway H, Stansfeld S, Marmot M. 2003. Is retirement good or bad for mental and physical health functioning? Whitehall II longitudinal study of civil servants. *J Epidemiol Community Health* 57: 46-49.

Melhorn JM. 2005. Working with common upper extremity problems. In *A Physician's Guide to Return to Work* (Ed. Talmage JB, Melhorn JM) : 181-213, American Medical Association, Chicago.

Merz MA, Bricout JC, Koch LC. 2001. Disability and job stress: implications for vocational rehabilitation planning. *Work* 17: 85-95.

Michie S, Williams S. 2003. Reducing work related psychological ill health and sickness absence: a systematic literature review. *Occup Environ Med* 60: 3-9.

Mital A, Mital A. 2002. Returning coronary heart disease patients to work: A modified perspective. *Journal of Occupational Rehabilitation* 12: 31-42.

Monninkhof EM, van der Valk PDLPM, van der Palen J, van Herwaarden CLA, Partridge MR, Walters EH, Zielhuis GA. 2003. Self-management education for chronic obstructive pulmonary disease (Cochrane Review). In *The Cochrane Library, Issue 4* John WIley & Sons, Ltd, Chichester.

Mookadam F, Arthur HM. 2004. Social support and its relationship to morbidity and mortality after acute myocardial infarction. *Arch Intern Med* 164: 1514-1518.

Morrell S, Taylor R, Quine S, Kerr C, Western J. 1994. A cohort study of unemployment as a cause of psychological disturbance in Australian youth. *Soc Sci Med* 38: 1553-1564.

Morrell SL, Taylor RJ, Kerr CB. 1998. Unemployment and young people's health. *MJA* 168: 236-240.

Morris JK, Cook DG, Shaper AG. 1992. Non-employment and changes in smoking, drinking, and body weight. *BMJ* 304: 536-541.

Morris JK, Cook DG, Shaper AG. 1994. Loss of employment and mortality. *BMJ* 308: 1135-1139.

Motor Accidents Authority. 2001. *Guidelines for the management of whiplash-associated disorders.* Motor Accidents Authority, Sydney.

Mowlam A, Lewis J. 2005. *Exploring how general practitioners work with patients on sick leave. A study commissioned as part of the 'Job retention and rehabilitation pilot' evaluation (DWP Research Report No 257).* Corporate Document Services, Leeds. **www.dwp.gov.uk/asd/asd5/rports2005-2006/rrep257.pdf** (accessed 9 December 2005).

Moylan S, Millar J, Davies R. 1984. For richer, for poorer? DHSS cohort study of unemployed men. DHSS Social Research Branch. Research Report No 11. Her Majesty's Stationery Office, London.

Murphy GC, Athanasou JA. 1999. The effect of unemployment on mental health. *Journal of Occupational and Organizational Psycholoy* 72: 83-99.

National Health and Medical Research Council. 2004. *Evidence-based management of acute musculoskeletal pain: a guide for clinicians.* Australian Academic Press pty ltd, Bowen Hills,Queensland.

National Research Council. 1999. *Work-related musculoskeletal disorders: report, workshop summary and workshop papers.* National Academy Press, Washington DC.

National Research Council. 2001. *Musculoskeletal disorders and the workplace.* National Academy Press, Washington DC.

Nelson DL, Simmons BL. 2003. Health psychology and work stress: a more positive approach. In *Handbook of occupational health psychology* (Ed. Quick JC, Tetrick LE) : 97-119, American Psychological Association, Washington DC.

Newman S. 2004. Engaging patients in managing their cardiovascular health. *Heart* 90(Suppl IV): iv9-iv13.

NHS CRD. 1998. Cardiac rehabilitation. *Effective Health Care Bulletin* 4 (NHS Centre for Reviews and Dissemination **www.york.ac.uk/inst/crd/ehc44.pdf**) (accessed 18 January 2006): 1-12.

Nicholson PJ, Cullinan P, Newman Taylor AJ, Burge PS, Boyle C. 2005. Evidence based guidelines for the prevention, identification, and management of occupational asthma. *Occup Environ Med* 62: 290-299.

NIDMAR. 2004. *Code of practice for disability management: describing effective benchmarks for the creation of workplace-based disability management programs.* National Institute of Disability Management and Research, Victoria, Canada.

NIOSH. 1997. *Musculoskeletal disorders and workplace factors: a critical review of epidemiologic evidence for work-related musculoskeletal disorders of the neck, upper extremity, and low back.* National Intstitute for Occupational Safety and Health, Cincinnati.

Nordenmark M. 1999. Non-financial employment motivation and well-being in different labour market situations: a longitudinal study. *Work, Employment and Society* 13: 601-620.

Nordenmark M, Strandh M. 1999. Towards a sociological understanding of mental well-being among the unemployed: the role of economic and psychosocial factors. *Sociology* 33: 577-597.

O'Brien GE, Feather NT. 1990. The relative effects of unemployment and quality of employment on the affect, work values and personal control of adolescents. *J Occup Psychol* 63: 151-165.

OECD. 2003. *Transforming disability into ability. Policies to promote work and income security for disabled people.* The Organisation for Economic Co-operation and Development, Paris.

Oliver M, Barnes C. 1998. *Disabled people and social policy: from exclusion to inclusion.* Longman, London & New York.

Ostry AS, Barroetavena M, Hershler R, Kelly S, Demers PA, Teschke K, Hertzman D. 2002. Effect of de-industrialisation on working conditions and self reported health in a sample of manufacturing workers. *J Epidemiol Community Health* 56: 506-509.

Palmore A. 2006. *The truth about stress.* Atlantic Books, London.

Pattani S, Constantinovici N, Williams S. 2004. Predictors of re-employment and quality of life in NHS staff one year after early retirement because of ill health; a national prospective study. *Occup Environ Med* 61: 572-576.

Patton W, Noller P. 1984. Unemployment and youth: a longitudinal study. *Australian J Psychology* 36: 399-413.

Patton W, Noller P. 1990. Adolescent self-concept: effects of being employed, unemployed or returning to school. *Australian J Psychology* 42: 247-259.

Payne R. 1999. Stress at work: a conceptual framework. In *Stress in health professionals. Psychological and organisational causes and interventions* (Ed. Firth-Cozens J, Payne RL) : 3-16, John WIley & Sons Ltd, Chichester.

Payne R, Jones JG. 1987. Social class and re-employment: changes in health and perceived financial circumstances. *Journal of Occupational Behaviour* 8: 175-184.

Perk J, Alexanderson K. 2004. Sick leave due to coronary artery disease or stroke. *Scand J Public Health* 32 (Suppl 63): 181-206.

Pickering T. 1997. The effects of occupational stress on blood pressure in men and women. *Acta Physiol Scand Suppl* 640: 125-128.

Platt S. 1984. Unemployment and suicidal behaviour: a review of the literature. *Soc Sci Med* 19: 93-115.

Poissonnet CM, Véron M. 2000. Health effects of work schedules in healthcare professions. *Journal of Clinical Nursing* 9: 13-23.

Proper KI, Koning M, van der Beek AJ, Hildebrandt VH, Bosscher RJ, van Mechelen W. 2003. The effectiveness of worksite physical activity programs on physical activity, physical fitness, and health. *Clin J Sport Med* 13: 106-117.

Proudfoot J, Guest D, Carson J, Dunn G, Gray J. 1997. Effect of cognitive-behavioural training on job-finding among long-term unemployed people. *The Lancet* 350: 96-100.

Punnett L, Wegman DH. 2004. Work-related musculoskeletal disorders: the epidemiologic evidence and the debate. *J Electromyogr Kinesiol* 14: 13-23.

Quaade T, Engholm G, Johansen AMT, Møller H. 2002. Mortality in relation to early retirement in Denmark: a population-based study. *Scand J Public Health* 30: 216-222.

Quinlan M, Mayhew C, Bohle P. 2001. The global expansion of precarious employment, work disorganization, and consequences for occupational health: a review of recent research. *International Journal of Health Services* 31: 335-414.

Rawls J. 1999. *A theory of justice.* Oxford University Press, Oxford.

RCGP. 1999. *Clinical Guidelines for the Management of Acute Low Back Pain.* Royal College of General Practitioners, London.

RCP. 2002. *Employment opportunities and psychiatric disability. Council Report CR111.* Royal College of Psychiatrists, London.

Reitzes DC, Mutran EJ, Fernandez ME. 1996. Does retirement hurt well-being? Factors influencing self-esteem and depression among retirees and workers. *The Gerontologist* 36: 649-656.

Reynolds MW, Frame D, Scheye R, Rose ME, George S, Watson JB, Hlatky MA. 2004. A systematic review of the economic burden of chronic angina. *Am J Manag Care* 10(11 Suppl): S347-S357.

Rhoades L, Eisenberger R. 2002. Perceived organizational support: a review of the literature. *Journal of Applied Psychology* 87: 698-714.

Rick J, Briner RB. 2000. Psychosocial risk assessment: problems and prospects. *Occupational Medicine* 50: 310-314.

Rick J, Briner RB, Daniels K, Perryman S, Guppy A. 2001. *A critical review of psychosocial hazard measures (HSE CRR 356)*. HSE Books, London.

Rick J, Thomson L, Briner RB, O'Regan S, Daniels K. 2002. *Review of existing supporting scientific knowledge to underpin standards of good practice for key work-related stressors - Phase 1. (HSE RR 024)*. HSE Books, London.

Ritchie H, Casebourne J, Rick J. 2005. *Understanding workless people and communities: a literature review (DWP Research Report No 255)*. Corporate Document Services, Leeds.

Rosenheck RA, Dausey DJ, Frisman L, Kasprow W. 2000. Outcomes after initial receipt of social security benefits among homeless veterans with mental illness. *Psychiatr Serv* 51: 1549-1554.

Rowlingson K, Berthoud R. 1996. *Disability, benefits and employment (DSS Research Report No 54)*. The Stationery Office, London.

Ryff CD, Singer B. 1998. The contours of positive human health. *Psychological Inquiry* 9: 1-28.

Salokangas RKR, Joukamaa M. 1991. Physical and mental health changes in retirement age. *Psychother Psychosom* 55: 100-107.

Salovey P, Rothman AJ, Detweiler JB, Steward WT. 2000. Emotional states and physical health. *American Psychologist* 55: 110-121.

Saunders P. 2002a. Mutual obligation, participation and popularity. Social security reform in Australia. *Journal of Social Policy* 31: 21-38.

Saunders P. 2002b. *The direct and indirect effects of unemployment on poverty and inequality. (SPRC discussion paper 118)*. Social Policy Research Centre, Sydney. **www.sprc.unsw.edu.au/dp/DP118.pdf** (accessed 09 December 2005).

Saunders P, Taylor P. 2002. *The price of prosperity. The economic and social costs of unemployment*. University of New South Wales Press, Sydney.

Scales J, Scase R. 2000. *Fit and fifty? A report prepared for the Economic and Social Research Council.* Insitute for Social and Economic Research, Universiy of Eessex and University of Kent at Canterbury, **www.esrc.ac.uk/ESRCInfoCentre/Images/Fit%20and%20Fifty_tcm6-6054.pdf** (accessed 29 November 2005).

Schaufeli W, Enzmann D. 1998. *The burnout companion to study and practice: A critical analysis.* Taylor & Francis, London.

Schaufeli WB. 1997. Youth unemployment and mental health: some Dutch findings. *Journal of Adolescence* 20: 281-292.

Scheel IB, Hagen KB, Herrin J, Carling C, Oxman AD. 2002a. Blind faith? The effects of promoting active sick leave for back pain patients: a cluster-randomized controlled trial. *Spine* 27: 2734-2740.

Scheel IB, Hagen KB, Herrin J, Oxman AD. 2002b. A randomized controlled trial of two strategies to implement active sick leave for patients with low back pain. *Spine* 27: 561-566.

Scheel IB, Hagen KB, Oxman AD. 2002c. Active sick leave for patients with back pain: All the players onside, but still no action. *Spine* 27: 654-659.

Schnall PL, Landsbergis PA, Baker D. 1994. Job strain and cardiovascular disease. *Annu Rev Public Health* 15: 381-411.

Schneider J. 1998. Work interventions in mental health care: some arguments and recent evidence. *J Mental Health* 7: 81-94.

Schneider J, Heyman A, Turton N. 2002. *Occupational outcomes: from evidence to implementation. (An expert topic paper commissioned by the Department of Health).* Centre for Applied Social Studies, University of Durham, Durham.

Schneider J, Heyman A, Turton N. 2003. *Employment for people with mental health problems: Expert briefing.* National Institute for Mental Health in England, **www.nimhe.org.uk/whatshapp/item_display_publications.asp?id=324**

Schonstein E, Kenny J, Keating J, Koes BW. 2003. Work conditioning, work hardening and functional restoration for workers with back and neck pain (Cochrane Review). In *The Cochrane Library, Issue 3* Update Software, Oxford.

Schwefel D. 1986. Unemployment, health and health services in German-speaking countries. *Sco Sci Med* 22: 409-430.

Scottish Executive. 2004. *Healthy working lives: a plan for action. Strategy paper.* Scottish Executive, Edinburgh.

Scottish Executive. 2005. *Social focus on deprived areas 2005*. National Statistics Publication, Edinburgh **www.scotland.gov.uk/Publications/2005** (accessed 21 March 2006).

Seymour L, Grove B. 2005. *Workplace interventions for people with common mental health problems: evidence review and recommendations*. British Occupational Health Research Foundation, London.

Shah H, Marks N. 2004. *A well-being manifesto for a flourishing society*. New Economic Foundation, London.

Shanfield SB. 1990. Return to work after an acute myocardial infarction: a review. *Heart Lung* 19: 109-117.

Shephard RJ. 1999. Age and physical work capacity. *Experimental Aging Research* 25: 331-343.

Sherrer YS. 2005. Working with common rheumatologic disorders. In *A Physician's Guide to Return to Work* (Ed. Talmage JB, Melhorn JM) : 289-304, American Medical Association, Chicago.

Shortt SED. 1996. Is unemployment pathogenic? A review of current concepts with lessons for policy planners. *International Journal of Health Services* 26: 569-589.

Simon GE, Barber C, Birnbaum HG, Frank RG, Greenberg PE, Rose RM, Wang PS, Kessler RC. 2001. Depression and work productivity: the comparative costs of treatment versus non-treatment. *J Occup Environ Med* 43: 2-9.

Slavin R. 1995. Best evidence synthesis: an intelligent alternative to meta-analysis. *J Clin Epidemiol* 48: 9-18.

Smeaton D, McKay S. 2003. *Working after state pension age: quantitative analysis (DWP Research Report No 182)*. Corporate Document Services, Leeds.

Smith L, Folkard S, Tucker P, Macdonald I. 1998. Work shift duration: a review comparing eight hour and 12 hour shift systems. *Occup Environ Med* 55: 217-229.

Smith R. 1985. Occupationless health. *British Medical Journal* 291: 1024-1027.

Smith R. 1987. *Unemployment and health: a disaster and a challange*. Oxford University Press, Oxford.

Snashall D. 2003. Hazards of work. In *ABC of occupational and environmental medicine* (Ed. Snashall D, Patel D) : 1-5, BMJ Books, London.

Sparks K, Cooper C, Fried Y, Shirom A. 1997. The effects of hours of work on health: a meta-analytic review. *Journal of Occupational and Organizational Psychology* 70: 391-408.

Sparks K, Faragher B, Cooper CL. 2001. Well-being and occupational health in the 21st century workplace. *Journal of Occupational and Organizational Psychology* 74: 489-509.

Spitzer WO, Skovron ML, Salmi LR, Cassidy JD, Duranceau J, Suissa S, Zeiss E. 1995. Scientific monograph of the Quebec Task Force on whiplash-associated disorders: redefining "whiplash" and its management. *Spine* 20, Supplement: 8S-73S.

Staal JB, Hlobil H, van Tulder MW, Waddell G, Burton AK, Koes BW, van Mechelen W. 2003. Occupational health guidelines for the management of low back pain: an international comparison. *Occup Environ Med* 60: 618-626.

Staal JB, Rainville J, Fritz J, van Mechelen W, Pransky G. 2005. Physical exercise interventions to improve disability and return to work in low back pain: current insights and opportunities for improvement. *Journal of Occupational Rehabilitation* 15: 491-502.

Steingrimsdottir OA, Vollestad NK, Knardahl S. 2004. Variation in reporting of pain and other subjective health complaints in a working population and limitations of single sample measurements. *Pain* 110: 130-139.

Svensson PG. 1987. International social and health policies to prevent ill health in the unemployed: The World Health Organization perspective. *Social Science & Medicine* 25: 201-204.

Sverke M, Hellgren J, Näswall K. 2002. No security: A meta-analysis and review of job insecurity and its consequences. *Journal of Occupational Health Psychology* 7: 242-264.

Talmage JB, Melhorn JM. 2005. *A physician's guide to return to work*. American Medical Association, Chicago.

Tarlo SM, Liss GM. 2005. Prevention of occupational asthma - Practical implications for occupational physicians. *Occupational Medicine* 55: 588-594.

The Council of the European Union. 2003. Council resolution of 15 July 2003 on promoting the employment and social integration of people with disabilities. *Official Journal of the European Union* OJ C175, 24.7.2003: 1-2.

The WHOQOL Group. 1995. The World Health Organization quality of life assessment (WHOQOL): position paper from the World Health Organization. *Soc Sci Med* 41: 1403-1409.

Thomas C, Benzeval M, Stansfeld SA. 2005. Employment transitions and mental health: an analysis from the British household panel survey. *J Epidemiol Community Health* 59: 243-249.

Thomas T, Secker J, Grove B. 2002. *Job retention and mental health: a review of the literature*. King's College London, London.

Thompson DR, Bowman GS, Kitson AL, de Bono DP, Hopkins A. 1996. Cardiac rehabilitation in the United Kingdom: guidelines and audit standards. *Heart* 75: 89-93.

Thompson DR, Lewin RJP. 2000. Management of the post-myocardial infarction patient: rehabilitation and cardiac neurosis. *Heart* 84: 101-105.

Tiggemann M, Winefield AH. 1984. The effects of unemployment on the mood, self-esteem, locus of control, and depressive affect of school-leavers. *Journal of Occupational Psychology* 57: 33-42.

Tsai SP, Wendt JK, Donnelly RP, de Jong G, Ahmed FS. 2005. Age at retirement and long term survival of an industrial population: prospective cohort study. *BMJ* doi:10.1136/bmj.38586.448704.EO.

Tsutsumi A, Kawakami N. 2004. A review of empirical studies on the model of effort-reward imbalance at work: reducing occupational stress by implementing a new theory. *Social Science & Medicine* 59: 2335-2359.

TUC. 2000. *Consultation document on rehabilitation. Getting better at getting back.* Trades Union Congress, London.

TUC. 2002. *Rehabilitation and retention - what works is what matters.* Trades Union Congress, London.

Tuomi K, Ilmarinen J, Seitsamo J, Huuhtanen P, Martikainen R, Nygård C-H, Klockars M. 1997. Summary of the Finnish research project (1981 - 1992) to promote the health and work ability of aging workers. *Scan J Work Environ Health* 23 (Suppl. 1): 66-71.

Tveito TH, Halvorsen A, Lauvålien JV, Eriksen HR. 2002. Room for everyone in working life? 10% of the employees - 82% of the sickness leave. *Norsk Epidemiologi* 12: 63-68.

Twamley EW, Jeste DV, Lehman AF. 2003. Vocational rehabilitation in schizophrenia and other psychotic disorders: a literature review and meta-analysis of randomized controlled trials. *The Journal of Nervous and Mental Disease* 191: 515-523.

United Nations. 1948. *Universal Declaration of Human Rights. United Nations General Assembly Resolution 217A(III).* United Nations, Geneva.

United Nations. 1975. *Declaration on the Rights of Disabled Persons. United Nations General Assembly Resolution 34/47.* United Nations, Geneva.

Ursin H. 1997. Sensitization, somatization, and subjective health complaints: a review. *Internat J Behav Med* 4: 105-116.

Ursin H, Eriksen HR. 2004. The cognitive activation theory of stress. *Psychoneuroendocrinology* 29: 567-592.

van der Doef M, Maes S. 1998. The job demand-control(-suppor) model and physical health outcomes: a review of the strain and buffer hypotheses. *Psychology and Health* 13: 909-936.

van der Doef M, Maes S. 1999. The job demand-control(-support) model and psychological well-being: a review of 20 years of empirical research. *Work & Stress* 13: 87-114.

van der Hulst M. 2003. Long work hours and health. *Scan J Work Environ Health* 29: 171-188.

van Dixhoorn J, White A. 2005. Relaxation therapy for rehabilitation and prevention in ischaemic heart disease: a systematic review and meta-analysis. *Eur J Cardiovasc Prev Rehhabil* 12: 193-202.

van Ryn M, Vinokur AD. 1992. How did it work? An examination of the mechanisms through which an intervention for the unemployed promoted job-search behavior. *American Journal of Community Psychology* 20: 577-597.

van Vegchel N, de Jonge J, Bosma H, Schaufeli W. 2005. Reviewing the effort-reward imbalance model: drawing up the balance of 45 empirical studies. *Social Science & Medicine* 60: 1117-1131.

Vingård E, Alexanderson K, Norlund A. 2004a. Consequences of being on sick leave. *Scand J Public Health* 32: 207-215.

Vingård E, Alexanderson K, Norlund A. 2004b. Sickness presence. *Scand J Public Health* 32: 216-221.

Vinokur A, Caplan RD. 1987. Attitudes and social support: determinants of job-seeking behavior and well-being among the unemployed. *Journal of Applied Social Psychology* 17: 1007-1024.

Vinokur A, Caplan RD, Williams CC. 1987. Effects of recent and past stress on mental health: coping with unemployment among Vietnam veterans and nonveterans. *Journal of Applied Social Psychology* 17: 710-730.

Vinokur AD, Price RH, Caplan RD. 1991a. From field experiments to program implementation: assessing the potential outcomes of an experimental intervention program for unemployed persons. *American Journal of Community Psychology* 19: 543-562.

Vinokur AD, Price RH, Schul Y. 1995. Impact of the JOBS intervention on unemployed workers varying in risk for depression. *American Journal of Community Psychology* 23: 39-74.

Vinokur AD, Schul Y. 1997. Mastery and inoculation against setbacks as active ingredients in the JOBS intervention for the unemployed. *J Consult Clin Psychol* 65: 867-877.

Vinokur AD, van Ryn M, Gramlich EH, Price RH. 1991b. Long-term follow-up and benefit-cost analysis of the JOBS project: a preventive intervention for the unemployed. *J Appl Psychol* 76: 213-219.

Vinokur AD, Vuori J, Schul Y, Price RH. 2000. Two years after a job loss: longterm impact of the JOBS Program on reemployment and mental health. *Journal of Occupational Health Psychology* 5: 32-47.

Virtanen P. 1993. Unemployment, re-employment and the use of primary health care services. *Scand J Primary Health Care* 11: 228-233.

Viswesvaran C, Sanchez JI, Fisher J. 1999. The role of social support in the process of work stress: a meta-analysis. *Journal of Vocational Behavior* 54: 314-334.

Vuori J, Silvonen J, Vinokur AD, Price RH. 2002. The Työhön Job Search Program in Finland: benefits for the unemployed with risk of depression or discouragement. *Journal of Occupational Health Psychology* 7: 5-19.

Vuori J, Vesalainen J. 1999. Labour market interventions as predictors of re-employment, job seeking activity and psychological distress among the unemployed. *Journal of Occupational and Organizational Psycholoy* 72: 523-538.

Waddell G. 2002. *Models of disability: using low back pain as an example.* Royal Society of Medicine Press, London.

Waddell G. 2004a. *Back Pain Revolution.* Churchill Livingstone, Edinburgh.

Waddell G. 2004b. *Compensation for chronic pain.* TSO, London.

Waddell G, Aylward M. 2005. *The scientific and conceptual basis of incapacity benefits.* The Stationery Office, London.

Waddell G, Burton AK. 2001. Occupational health guidelines for the management of low back pain at work: evidence review. *Occup Med* 51: 124-135.

Waddell G, Burton AK. 2004. *Concepts of rehabilitation for the management of common health problems.* The Stationery Office, London.

Wagstaff A. 1985. Time series analysis of the relationship between unemployment and mortality: a survey of econometric critiques and replications of Brenner's studies. *Soc Sci Med* 21: 985-996.

Wainwright D, Calnan M. 2002. *Work stress. The making of a modern epidemic.* Open University Press, Buckingham.

Walker-Bone K, Cooper C. 2005. Hard work never hurt anyone: or did it? A review of occupational associations with soft tissue musculoskeletal disorders of the neck and upper limb. *Ann Rheum Dis* 64: 1391-1396.

Wanberg CR. 1995. A longitudinal study of the effects of unemployment and quality of reemployment. *Journal of Vocational Behavior* 46: 40-54.

Warr P. 1987. *Work, unemployment, and mental health.* Oxford University Press, Oxford.

Warr P. 1994. A conceptual framework for the study of work and mental health. *Work & Stress* 8: 84-97.

Warr P, Butcher V, Robertson I, Callinan M. 2004. Older people's well-being as a function of employment, retirement, environmental characteristics and role preference. *British Journal of Psychology* 95: 297-324.

Warr P, Jackson P. 1985. Factors influencing the psychological impact of prolonged unemployment and of re-employment. *Psychological Medicine* 15: 795-807.

Watson PJ, Booker CK, Moores L, Main CJ. 2004. Returning the chronically unemployed with low back pain to employment. *European Journal of Pain* 8: 359-369.

Weber A, Lehnert G. 1997. Unemployment and cardiovascular diseases: a causal relationship? *Int Arch Occup Environ Health* 70: 153-160.

Wegman DH. 1999. Older workers. *Occupational Medicine: State of the Art Reviews* 14: 537-557.

Wenger NK, Froelicher ES, Smith LK. 1995. *Cardiac rehabilitation as secondary prevention. Clinical Practice Guideline. Quick Reference Guide for Clinicians, No 17.* Department of Health and Human Services, Public Health Service, Agency for Health Care Policy and Research and National Heart, Lung, and Blood Institute. AHCPR Publications No. 96-0672, Rockville.

Wessely S. 2004. Mental health issues. In *What about the workers? Proceedings of an RSM Symposium* (Ed. Holland-Elliot K) : 41-46, Royal Society of Medicine Press, London.

Westgaard RH, Winkel J. 1997. Ergonomic intervention research for improved musculoskeletal health: A critical review. *International Journal of Industrial Ergonomics* 20: 463-500.

Westin S. 1993. Does unemployment increase the use of primary health care services? (Editorial). *Scand J Primary Health Care* 11: 225-227.

Westman A, Linton SJ, Theorell T, Öhrvik J, Wahlén P, Leppert J. 2006. Quality of life and maintenance of improvements after early multimodal rehabilitation: A 5-year follow-up. *Disability and Rehabilitation* 28: 437-446.

White S. 2004. A social democratic framework for benefit conditionality. In *Sanctions and sweeteners: rights and responsibilities in the benefits system* (Ed. Stanley K, Lohde LA, White S) Institute for Public Policy Research, London.

WHO. 1948. *Preamble to the Constitution of the World Health Organisation.* World Health Organisation, Geneva **www.who.int/en** (accessed 10 April 2006).

WHO. 1995. *Global strategy on occupational health for all. The way to health at work (WHO/OCH/95.1)*. World Health Organisation, Geneva **www.who.int/occupational_health/globalstrategy/en/** (accessed 06 January 2006).

WHO. 2001. *Health and ageing - a discussion paper*. World Health Organisation, Geneva.

WHO. 2004. *Family of International Classifications: definition, scope and purpose*. World Health Organisation, Geneva **http://www.who.int/classifications/icd/docs/en/WHOFICFamily.pdf** (accessed 10 April 2006).

Williams R, Hill M, Davies R. 1999. *Attitudes to the welfare state and the response to reform. (DSS Research Report No. 88)*. Corporate Document Services, Leeds.

Winefield AH, Tiggemann M. 1990. Employment status and psychological well-being: a longitudinal study. *Journal of Applied Psychology* 75: 455-459.

Winefield AH, Tiggemann M, Winefield HR. 1990. Factors moderating the psychological impact of unemployment at different ages. *Person Individ Diff* 11: 45-52.

Winefield AH, Tiggemann M, Winefield HR. 1991a. The psychological impact of unemployment and unsatisfactory employment in young men and women: longitudinal and cross-sectional data. *Brit J Psychol* 82: 487-505.

Winefield AH, Winefield HR, Tiggemann M, Goldney RD. 1991b. A longitudinal study of the psychological effects of unemployment and unsatisfactory employment on young adults. *Journal of Applied Psychology* 76: 424-431.

Witt BJ, Thomas RJ, Roger VL. 2005. Cardiac rehabilitation after myocardial infarction: a review to understand barriers to participation and potential solutions. *Eura Medicophys* 41: 27-34.

Womack L. 2003. Cardiac rehabilitation secondary prevention programs. *Clin Sports Med* 22: 135-160.

Woods V. 2005. Work-related musculoskeletal health and social support. *Occupational Medicine* 55: 177-189.

Woods V, Buckle P. 2002. *Work, inequality and musculoskeletal health (Contract Research Report 421)*. Health and Safety Executive, London.

Wozniak MA, Kittner SJ. 2002. Return to work after ischemic stroke: a methodological review. *Neuroepidemiology* 21: 159-166.

Young A, Roessler RT, Wasiak R, McPherson KM, van Poppel MNM, Anema JR. 2005. A developmental conceptualization of return to work. *J Occup Rehabil* 15: 557-568.

Evidence tables

TABLE1. HEALTH EFFECTS OF WORK *vs* UNEMPLOYMENT

Authors	Key features (*Additional reviewers' comments in italics*)

Table 1a: Work

(Chirikos 1993) Review of economic studies	**The relationship between health and labour market status**

That the health of an individual will affect his or her labour market status and productivity seems self-evident based on a priori reasoning and casual observation. This review of US econometric studies from 1970-1990 provides unambiguous empirical proof that poorer health has an economic impact either by restricting physical or mental capacity to engage in work, lowering productivity, shifting economic choices about whether to work, or changing the labour market activity of other family members. However, there is considerable variation in the size of this effect across different studies. This is partly because of different methods of measuring the severity of the health condition and the impact on labour market activity, (*but also due to more fundamental conceptual difficulties about the relationship between health and work, and the extent to which individual decisions fit the model of 'the economic man'*)

(WHO 1995) World Health Organisation strategy	**Global Strategy on Occupational Health for All: The Way to Health at Work**

According to the principles of the United Nations, the World Health Organisation and the International Labour Organisation, every citizen of the world has a right to healthy and safe work and to a work environment that enables him or her to live a socially and economically productive life.

Conditions of work and the work environment may have either a positive or hazardous impact on health and well-being. Ability to participate in the working life opens the individual possibilities to carry out economically independent life, develop his or her working skills and social contacts. One-third of adult life is spent at work where the economic and material values of society are generated. On the other hand, dangerous exposures and loads are often several times greater in the workplace than in any other environment with adverse consequences on health.

Health at work and healthy work environments are among the most valuable assets of individuals, communities and countries. Occupational health is an important strategy not only to ensure the health of workers, but also to contribute positively to productivity, quality of products, work motivation, job satisfaction and thereby to the overall quality of life of individuals and society.

TABLE1. HEALTH EFFECTS OF WORK vs UNEMPLOYMENT

Authors	Key features (*Additional reviewers' comments in italics*)

Table 1 a: Work *continued*

(Harrington 1994a) Narrative review (Harrington 1994b) Editorial	**Shift work and health** (Narrative review but based on the findings of multiple, referenced studies). Working outside normal hours either by extended days or shift work can have detrimental health effects in the form of circadian rhythm disturbance, poorer quality and quantity of sleep and increased fatigue. There is a persuasive body of evidence that poorer work performance and output and increased accidents are associated with shift work, particularly on the night shift, though individual factors may be as important as workplace factors. The link between shift work and increased cardiovascular morbidity and mortality has strengthened in recent years. The case for an association with gastrointestinal disease remains quite good. Optimal hours for the working week cannot be formulated on present scientific evidence: there is no unequivocal scientific evidence to support the European directive for a 48 hour week, though there is limited evidence that working >48-56 hours per week may carry serious health and safety implications.
(Sparks *et al.* 1997) Systematic review meta-analysis	**The effect of hours of work on health** Meta-analysis of 21 studies showed weak but significant positive correlations (mean r = 0.13) between overall and health symptoms, physiological and psychological health symptoms and hours of work. Qualitative analysis of 12 additional studies supported these findings. Concluded that working long hours (possibly >48 hours/week, though that figure was based on limited evidence) can be detrimental to physiological and psychological health. However, many factors may have a moderating effect on the complex relationship between working hours and health, e.g. the nature of the job, the working environment, age, whether working long hours is a matter of choice, life style and the impact on health behaviours (e.g. smoking, eating habits and exercise). (*In 2004, average full-time employment in UK was 43.4 hours per week, compared with EU15 41.7 hours per week. UK was fourth highest after Iceland, Austria and Greece. Eurostat http://epp.eurostat.cec.eu.int accessed 08 November 2005*)

TABLE 1. HEALTH EFFECTS OF WORK vs UNEMPLOYMENT

Authors	Key features (*Additional reviewers' comments in italics*)

Table 1a: Work *continued*

(Smith *et al.* 1998) Systematic review	**Work shift duration: comparing 8 hour and 12 hour shifts** A compressed working week is defined as 'Any system of fixed working hours more than eight hours in duration which results in a working week of less than five full days of work a week'. This review considers the evidence on 8 and 12 hour shifts for: (a) fatigue and performance in work; (b) safety; (c) sleep, physical health, and psychosocial wellbeing; (d) system implementation, shiftworker attitudes, preferences, and morale; (e) absenteeism and turnover; and (f) overtime and moonlighting. The research findings are largely equivocal. The evidence on the effects of 12 hour systems on fatigue and job performance is equivocal. Provided adequate safety measures are taken, there seems to be no conclusive evidence that extended work shifts compromise safety from the point of view of increased accident rates, job performance, or increased error rates. However, fatigue and safety remain concerns in jobs with high workload and demands. Much of the evidence suggests that shiftworkers do not have great problems with sleep, health, and well-being when working 12 hour compared with 8 hour shifts and may even show improvements in these areas. In general, Depending on how 12 hour shifts are introduced, there can be high employee acceptance and satisfaction, self rated stress levels may be considerably reduced, while family and social life are improved. In general, absenteeism and turnover do not increase with 12 hour systems, though older workers may find it more difficult to deal with those rotas. The issue of additional overtime is also an area of some concern with 12 hour shift systems.
(Acheson *et al.* 1998) Independent Department of Health Inquiry	**Inequalities in health** Despite a marked increase in prosperity and substantial reductions in mortality for the UK population as a whole over the past 20+ years, the gap in health between those at the top and the bottom of the social scale has widened. These inequalities affect the whole of society and can be identified at all stages of the life course from pregnancy to old age. The weight of scientific evidence supports a socio-economic explanation of health inequalities. This traces the roots of ill health to such determinants as income, education and employment as well as to the material environment and lifestyle. The Report made a wide range of policy recommendations, the most relevant of which for the present review included: • further steps to reduce income inequalities and improve the living standards of poor households; • policies to improve the opportunities for work and to ameliorate the health consequences of unemployment; • policies to improve the quality of jobs, and to reduce psychosocial work hazards.

TABLE 1. HEALTH EFFECTS OF WORK vs UNEMPLOYMENT

Authors	Key features (*Additional reviewers' comments in italics*)

Table 1a: Work *continued*

Authors	Key features
(Ferrie 1999) Conceptual narrative review (Ferrie 2001) Brief narrative review.	**Health consequences of job insecurity** The most important element of job insecurity is the employee's perception that his or her job is not safe: the job insecurity experience then depends on the perceived probability of losing one's job and the perceived severity of the effects. There are several large categories of workers with insecure jobs, the most vulnerable of which include: a) those in the secondary labour market, e.g. foreign workers, immigrants, ethnic minorities, older workers and to some extent women, especially those with young children; b) workers on short-term contracts. 15 longitudinal studies of workplace closure nearly all showed one or more adverse effects on physical health, physiological indicators and psychological health during the anticipation and termination phases and the first year of unemployment. In general, health-related behaviour remained unchanged or improved. 5 studies showed that re-employment may partially or completely reverse these adverse effects.
(Baltes *et al.* 1999) Meta-analysis	**Flexible and compressed workweek schedules: their effects on work-related criteria.** In general, both flexible and compressed (into <5 days) workweek schedules had positive effects. Flexible work schedules produced significant improvements in absenteeism, productivity, job satisfaction and satisfaction with work schedule. However, less flexible schedules had larger effect sizes than highly flexible schedules, and the effects diminished over time. Compressed schedules produced significant improvements in supervisor performance rating, job satisfaction and satisfaction with work schedule but did not affect absenteeism and productivity. Flextime had greater effects on behavioural than on attitudinal outcomes; compressed schedules had greater effects on attitudinal than on behavioural outcomes.
(Poissonnet & Véron 2000) Systematic review	**Health effects of work schedules in healthcare professions** There is no clear evidence to establish a link between irregular hours and physical health, gastrointestinal disorders or cardiovascular disease. Fatigue, nervousness, stress and burnout are commonly reported by health care professionals, but there is limited evidence to link this to work schedules. Sleep disorders are frequently associated with working shifts, particularly irregular shifts, though this varies with the individual. No conclusive evidence was found to favour any particular work schedule, though there is evidence that 12-hour shifts have negative effects on nurses' job performance, delivery of patient care, safety, job satisfaction and psychosocial well-being.

TABLE1. HEALTH EFFECTS OF WORK vs UNEMPLOYMENT

Authors	Key features (*Additional reviewers' comments in italics*)

Table 1a: Work *continued*

(Benavides *et al.* 2000) 2nd European Survey on working conditions	**How do types of employment relate to health indicators?** *(Cross-sectional survey of 15 European countries including UK.)* Precarious employment (defined for this review as fixed term contracts and temporary work) accounts for 15% of paid employment. It commonly involves low income levels, poor physical working conditions (e.g. exposure to noise, vibration, hazardous material, or repetitive tasks), poor psychosocial working conditions, unstable jobs and low social or legal protection. Precarious employment is positively but weakly associated with fatigue (adjusted odds ratio (OR) 1.12), backache (adjusted OR 1.08) and muscular pains (adjusted OR 1.21). It is consistently associated with lower job satisfaction (adjusted OR 1.79) but negatively associated with stress and absenteeism (when compared with full-time permanent workers).
(Sparks *et al.* 2001) Narrative review	**Well-being and occupational health in the 21st century workplace** Considers the impact on employee well-being of major changes in the workplace over the past 40 years: increased female participation, globalisation, organisational restructuring and downsizing, changes in work contracts and scheduling, more short-term contracts and flexible working, and the growth of information technology. Reviews organisational psychology research into four areas and their impact on the psychological well-being of employees: job insecurity (mainly the perception rather than necessarily the reality, and particularly among white collar workers), work hours, control at work (again primarily perceived control, based on the Karasek model), and managerial style. *(This is a narrative review from a psychological perspective, focusing on potential psychological harm. The conclusions are broadly comparable to those of other reviews, but somewhat less critical and with no consideration of effect sizes.)* The final message of the paper is the importance of good communications between managers and employees, which will enhance the effectiveness of any interventions to improve employee well-being and organisational performance.
(Quinlan *et al.* 2001) Systematic review	**The consequences of precarious employment for occupational health and safety** Precarious employment includes temporary/short term contracts, job insecurity associated with organisational restructuring, outsourced/home working, part-time work, and work in small businesses. Reviewed 93 studies, of which 76 found a negative association between precarious work and occupational health and safety in terms of hazard exposures, injury rates, disease risk or worker and manager knowledge of OHS and regulatory requirements. 11 studies had inconclusive findings, and only 6 found no relationship or produced contrary results. There was strong evidence for outsourcing and organizational restructuring/downsizing and mixed evidence for temporary work and small businesses. There was strong evidence that part-time work was not associated with adverse occupational health and safety outcomes.

TABLE1. HEALTH EFFECTS OF WORK *vs* UNEMPLOYMENT

Authors	Key features (*Additional reviewers' comments in italics*)

Table 1a: Work *continued*

(Sverke *et al.* 2002) Meta-analysis	**Job insecurity and its consequences** Job insecurity was defined as 'the subjectively experienced threat of involuntary job loss'. Reviewed reports of 86 independent samples including 38,531 individuals and rated both the strength of the evidence and the strength of the effects. There was strong evidence for a moderate negative association between job insecurity and mental health (corrected r = -0.24) and strong evidence for a weak negative association between job insecurity and self-reported general health (corrected r = -0.16). Regarding attitudes to work, there was strong evidence of a strong negative association between job insecurity and job satisfaction (corrected r = -0.41), and moderate evidence of a moderate negative association between job insecurity and job involvement (corrected r = -0.37). Regarding attitudes to the organisation, there was strong evidence of a strong negative association between job insecurity and trust (corrected r = - 0.50), and a moderate negative association between job insecurity and organisational commitment (corrected r = - 0.36). There was strong evidence that there was no significant association between job insecurity and work performance. There was strong evidence for a moderate association between job insecurity and turnover intention (corrected r = -0.28). Variation of findings between studies suggests that the relationships between job insecurity and the outcome variables are moderated by other factors. In particular, stronger stress reactions can be expected among less-favoured occupational groups with lower education and stronger economic dependence on paid employment. The major limitation of this evidence is that it is mainly based on correlational studies, though a few longitudinal studies provide limited evidence for a causal relationship.

TABLE1. HEALTH EFFECTS OF WORK vs UNEMPLOYMENT

Authors	Key features *(Additional reviewers' comments in italics)*
Table 1a: Work *continued*	
(Green 2003) Economic review	**The rise and fall of job insecurity**
	Job security – confidence in the continuity and progress of employment – is a core element in the quality of work. Unemployment has a deep and long-lasting negative impact on the well-being of workers. However, unemployment does not only affect those who actually experience it: job insecurity is the threat of unemployment for those in work. The most important elements of job insecurity are the risk and cost of job loss and the ease of re-employment. Job insecurity is subjective, a matter of perceptions and expectations, about uncertainty and ambiguity. Job insecurity generates severe psychological strains, is a substantive source of ill health and job dissatisfaction, and has long-lasting deleterious impacts on individuals and their families. In some studies, the effects of insecurity on well-being are found to be as great as or exceeding the impact of becoming unemployed.
	Historically, job insecurity was mainly a concern of blue collar workers. In the mid-1990s, it became a public and political issue, driven by its impact (probably for the first time) on white collar workers. Survey evidence suggests that job insecurity generally increased in the 1970s and 1980s. Contrary to popular belief, from the late 1990s, most occupation groups in Britain have experienced less insecurity, reflecting a return to historically low levels of unemployment. Today, insecure workers are concentrated in jobs with temporary or short contracts, the private sector, and foreign-owned firms.
(Dooley 2003) Narrative & conceptual review	**Unemployment, underemployment, and mental health: conceptualizing employment status as a continuum**
	Underemployment is defined as including both unemployment and economically inadequate types of employment. Low hours, low pay, unemployment and discouraged worker rates tend to go together. Research focusing on unemployment ignores the impact of other employment statuses. Dooley suggests that it is better to think of a continuum between working and nonworking, including more and less adequate types of employment. This review found relatively few studies of the health effects of underemployment and even fewer with longitudinal designs. It concluded that there is limited evidence that economically inadequate employment is associated with adverse mental health effects similar to those of job loss.

TABLE1. HEALTH EFFECTS OF WORK vs UNEMPLOYMENT

Authors | **Key features** (*Additional reviewers' comments in italics*)

Table 1a: Work *continued*

(van der Hulst 2003)
Systematic review

Long work hours and health

Earlier reviews (e.g. Sparks 1997) did not fully distinguish long working hours from possible confounders such as shift work and high job demands. This review included 27 empirical studies (20 cross-sectional) published 1996-2001, which controlled for possible covariates. Van der Hulst concluded that these showed that a) long working hours are associated with adverse health as measured by cardiovascular disease, diabetes, disability retirement, subjectively reported physical health, subjective fatigue; and b) that some evidence exists for an association between long work hours and physiological changes, e.g. cardiovascular and immunological parameters.

(*van der Hulst took any positive findings in >50% of studies on a given topic as 'positive' evidence, even if there were only a few studies and/or mixed findings. Using the present review's evidence rating system, the studies reviewed by van der Hulst provide limited and/or conflicting evidence that there may be a weak association between working long hours and raised cardiovascular or other physiological measures, increased risk of cardiovascular disease, and all-cause mortality. They provide conflicting evidence that working long hours has any effect on self-reported general health, physical health or psychological health.*)

TABLE1. HEALTH EFFECTS OF WORK vs UNEMPLOYMENT

Authors

Key features (*Additional reviewers' comments in italics*)

Table 1a: Work *continued*

(Coggon 1994)
(Snashall 2003)
Occupational health texts

Epidemiology of occupational disease
Hazards of work

Some (aspects of) work can be hazardous to health. 'Good' work is life enhancing, but bad working conditions can damage your health. Epidemiology, when applied to occupational morbidity lie in three areas: (1) identification of occupational hazards; (2) assessment of the risks from the hazards; (3) monitoring outcomes of measures to control health risks in the workplace. The effects of some occupational hazards are readily identified (e.g. inhalation of sulphur dioxide causing respiratory irritation) and some will be readily associated with specific occupational groups (e.g. asbestosis in laggers). However, the relation of hazard to effect is often less obvious. The disorder concerned may have other causes which make it common even in those not exposed to the occupational hazard (e.g. back pain in construction work). Common health problems for which occupational exposures are important but not the major cause may nonetheless interfere with work, or work may be more challenging in the face of the condition – these can be termed 'work-related' or 'work-relevant' conditions. Non-occupational factors related to behaviour and life-style may also be important, in addition to social disadvantage. Control of a hazard requires knowledge of the level of exposure related to the risk, but exposure-response relations are difficult to determine for many common health problems. Similarly, monitoring the control measures is problematic when the condition is loosely linked to occupation (e.g. because of inadequate reporting systems or misdiagnoses). Nevertheless, the process of hazard identification and risk control forms the backbone of much health and safety legislation and practice.

TABLE 1. HEALTH EFFECTS OF WORK vs UNEMPLOYMENT

Authors	Key features (*Additional reviewers' comments in italics*)

Table 1a: Work *continued*

(HSC 2002) (HSC 2004) UK policy documents	**The health and safety system in Great Britain** **A strategy for workplace health and safety in Great Britain** In Great Britain, risks to health and safety arising from work are regulated through a single legal framework whereby the Health and Safety Commission (HSC) (www.hse.gov.uk/aboutus/hsc/) and the Health and Safety Executive (HSE) (**www.hse.gov.uk**) have specific statutory functions. HSC's statutory responsibility under the Health and Safety at Work Act 1974 (HSW) include proposing health and safety law. In preparing its proposals, the HSC relies on advice from HSE based on HSE research activities. The standards of health and safety are delivered by the HSW and are typified by the Management of Health and Safety at Work Regulations 1992, which reflect a long tradition of health and safety regulation going back to the 19th century, and also relate to the EU Council Directive (89/391) 1989 (EU 1989). In essence, the system is one of regulation, assessment/control and enforcement: employers are required to assess and control (so far as is reasonably practicable) the health and safety risks from the work, submit to official inspections and face legal sanction if they fail to reduce any risks. This regime has cut workplace fatalities by around two-thirds since the 1970s, but these traditional interventions are less effective when dealing with health than when dealing with safety. The current strategy, whilst still focused on risk assessment, control and enforcement, encourages the contribution of all the stakeholders in the control of health and safety—there is a need for education, rehabilitation, and public service reform. Leverage on health issues will require new methods, including simplifying the process of risk assessment and strengthening the role of health and safety in getting people back to work. Through the Council Directive of 1989 and subsequent directives, the member states of the European Union have similar health and safety legislation and procedures guided by the European Agency for Safety and Health at Work (http://agency.osha.eu.int/OSHA), whilst the USA has comparable systems under the National Institute for Occupational Safety and Health (www.niosh.com) and the Occupational Safety and Health Administration (www.osha.gov). In UK some conditions are sufficiently closely associated with work exposure to be prescribed as industrial diseases or injuries by the Industrial Injuries Advisory Council (www.iiac.org.uk), though individual cases are still adjudicated. Whilst most developed countries have some form of health and safety legislation, there are many developing countries where the burden of occupational injury and disease is particularly heavy and uncontrolled, and environmental exposures can be an additional risk to health (Snashall 2003 - see below). (*There is widespread acceptance that certain aspects of work need to be controlled to ensure workers' safety and health. The assessment and control of risks works very well for prevention of serious accidents and exposure to substances hazardous to health, but it is far less effective for prevention of ill health, particularly in respect of common health problems. Nevertheless, certain aspects of work may be difficult for people with health complaints (irrespective of cause) or may aggravate symptoms and make return to work problematic; that is not the same as primary causation*).

TABLE 1. HEALTH EFFECTS OF WORK vs UNEMPLOYMENT

Authors	Key features (Additional reviewers' comments in italics)

Table 1a: Work continued

Authors	Key features
(HDA 2004) Briefing note	**The evidence about work and health** The broader relationship between work and health may be understood in terms of three mechanisms. First, work that provides for degrees of fulfilment or job satisfaction and particularly allows individuals discretion and control over their working lives appears to confer considerable health benefit when measured in terms of overall mortality. Second, conversely, jobs that are lacking in self-direction and control appear to confer far fewer health benefits and the rates of mortality and morbidity among workers appear to be consistently higher. Third, the absence of work (unemployment) produces considerable negative health effects. There is considerable debate about the precise ways in which these mechanisms work. (*These appear to be effects rather than 'mechanisms'*).
(Cox *et al.* 2004) Policy paper	**Work, employment and mental health in Europe** Occupational health is concerned with understanding the dynamic relationship between work on the one hand and health on the other, and with protecting and promoting health by exploiting this relationship. This model of occupational health includes both the effects of work on health and those of health on availability and fitness for work and on work ability. On the one hand, work can cause, contribute to or aggravate mental ill health. However, for those involuntarily out of work, employment and work can improve mental health and work may be therapeutic. On the other hand, for those in work, mental ill health can be a determinate or correlate of poor organisational performance, absenteeism or adverse health behaviours. It can delay, reduce the likelihood of, or prevent return to work. For those out of work, mental ill health can be an obstacle to job-seeking, gaining and sustaining employment, and can reduce the likelihood of, or prevent, re-employment. The relationship between work and mental health may be modulated by employment or work status, and by individual diversity such as gender, age, ethnicity and cultural background. Concludes by setting up a dichotomy between 'for working populations, work can be a major challenge to mental health, while for populations out-of-work it may be a pro-health factor'. (*This ignores the possibilities that work may have positive effects on the mental health of workers, and that worklessness may also sometimes have positive effects on mental health*).

TABLE 1. HEALTH EFFECTS OF WORK vs UNEMPLOYMENT

Authors	Key features (*Additional reviewers' comments in italics*)

Table 1a: Work *continued*

(Shah & Marks 2004) Manifesto	**A well-being manifesto for a flourishing society** (New Economics Foundation)

One of the key aims of a democratic government is to promote the good life: a flourishing society where citizens are happy, healthy, capable, and engaged – in other words with high levels of well-being. The dimensions of well-being include: a) satisfaction with life, which includes satisfaction, happiness and enjoyment; b) personal development, which includes engagement, personal growth and development, fulfilling individual potential, and having purpose and meaning in life; and c) social well-being – a sense of belonging to our communities, a positive attitude towards others, feelings of engaging in and contributing to society.

(*Most of the paper is devoted to the general social structure, social policy and what Government can do to improve life*). Suggests that the health care system needs to be re-focused to promote complete health. There are important links between health and well-being. The scale of the effect of psychological well-being on health is of the same order as traditionally identified risks such as body mass, lack of exercise, and smoking. The National Health Service and other health institutions need to continue to broaden their focus to promote complete health. To do this, we need to accelerate the move towards a preventative health system. We also need to tackle mental health far more systematically. Treating people holistically means that health professionals need to go beyond just curing the biomedical causes of disease to thinking about the social and psychological aspects of how patients are treated.

TABLE1. HEALTH EFFECTS OF WORK vs UNEMPLOYMENT

Authors	Key features (*Additional reviewers' comments in italics*)

Table 1a: Work *continued*

Authors	Key features
(Klumb & Lampert 2004) Systematic review	**Women, work and well-being** Review of research on the impact of employment on women's physical and mental health, including 140 studies (though only 13 were methodologically-sound, longitudinal studies with multivariate analysis controlling for covariates and providing effect sizes). Conceptually, there are two competing models about the impact of multiple social roles (i.e. adding employment to family roles): stress vs. enhancement hypotheses, implying broadly –ve or +ve impacts of employment on health respectively. The further question may be under what conditions employment has adverse or beneficial effects. The main conclusion of this review was that paid employment has either beneficial or neutral effects and, importantly, has no adverse effects on women's health. More specifically: 1) there was strong evidence that employment is associated with reduced psychological distress; 2) the majority of cross-sectional studies found that employment is associated with better general physical health, but the higher quality studies showed conflicting results (importantly, however, only 1 higher quality study showed any adverse effects); 3) there is conflicting evidence on whether employment had a beneficial effect on cardiovascular risk factors and disease, the majority of studies showing no effect (importantly, however, no higher quality study showed any adverse effects); 4) there is strong evidence that employment is associated with lower mortality, though the relative importance of selection and causal explanations remains unclear. These results support the enhancement model and are directly contrary to the occupational stress hypothesis.
(Layard 2004) Policy paper	**Good jobs and bad jobs** (Centre for Economic Performance) Argues that human happiness is more affected by whether or not one has a job than by what kind of job it is. Thus, when jobs are to hand, we should insist that unemployed people take them. This involves a much more pro-active placement service and clearer conditionality than applies in many countries.

TABLE1. HEALTH EFFECTS OF WORK vs UNEMPLOYMENT

Authors	Key features (*Additional reviewers' comments in italics*)

Table 1a: Work *continued*

(Dodu 2005)
Narrative review

Is employment good for well-being?

Employment is an integral and dominant part of life in the industrialised world, and has a complex relationship with individual well-being. Employment provides the individual with financial gains, social identity and status, a means of structuring time and a sense of personal achievement. These 'needs' improve self-esteem and contribute to an individual's sense of well-being. However, employment can also involve biopsychosocial stressors, which affect people through four inter-linked mechanisms – cognitive, emotional, behavioural and physiological – which vary with the individual and the context. Individuals interact with modern workplaces in complex, multifactorial ways that affect their health and well-being (*both positively and negatively*). Recognises that whilst unemployment is correlated with poor health, that does not prove the corollary that employment must therefore be good for health. Recognises a wide range of potential occupational physical and psychological risks to health, but also recognises aspects of work that may contribute to well-being. It is concluded that employment clearly contributes to the individual's role in society, enabling them to attain social status and psychological identity. Purposeful and meaningful activity, and a good work–life balance seem to be key. Being without employment is not always bad – keeping active and purposeful may modulate ill-effects of not working. Found there seems to be little research specifically on the positive association between employment and health. Concludes that the complex relationship between employment (and unemployment) and well-being is dependent on the individual and the situation – at best, when employment fulfils an individual's physical, social and psychological needs, it can contribute to well –being; however, at worst, work can make you ill. The answer to the question 'is work good for well-being?' is that 'it depends' – the challenge is how to assess the inherent complexities. (*This review focused on a very similar question to the present review and approached it, at least partly, from the perspective of occupational psychology. Deliberately set out to contribute to the debate and presented more questions than answers*).

(Deacon 2005)
Discussion paper

Health & Performance
(AXA/PPP Healthcare)

Employees are the most valuable asset of any organisation. 'Human capital' describes the potential value of employees to an organisation and includes their health, fitness, knowledge, skills, experience and well-being. 90% of companies recognise the strong link between employee health and productivity: productivity is not just a matter of employees' work capability but depends on their health, fitness, stamina and well-being. There is therefore a strong business case for investing in employees' health and well-being: 'better health means better business.' (*Discusses the organisation and provision of occupational health services, absenteeism, presenteeism, employee assistance programmes, employee health insurances, rehabilitation and the resulting productivity benefits*).

TABLE1. HEALTH EFFECTS OF WORK vs UNEMPLOYMENT

Authors	Key features *(Additional reviewers' comments in italics)*

Table 1a: Work *continued*

(Coats 2005) Policy paper	**An agenda for work: the Work Foundation's challenge to policy makers** (The Work Foundation) *(Challenges policy makers to address the whole range of issues relevant to the world of work). A clear account of what 'good work' comprises has yet to be articulated. The paper gives a vision of 'good work', the purpose not being to develop 'good work' for its own sake, but to recognise a need to apply a model in practice.* • To fill the gap in the national conversation policy makers must have a clear conception of what constitutes "good work". The factors that characterise "bad jobs" are well understood: • a lack of control over the pace of work and the key decisions that affect the workplace • limited task discretion and monotonous and repetitive work • Inadequate skill levels to cope with periods of intense pressure • an imbalance between effort and reward • limited "social capital" – whether informal friendship networks or formal associations like trade unions - which make workers more resilient. • A vision of "good work" should therefore embrace the following: • Full employment – defined as the availability of jobs for all those who wish to work • Fair pay (including equal pay for work of equal value) • The absence of discrimination on the grounds of race, gender, sexuality, disability or age • Secure and interesting jobs that employees find fulfilling • A style and ethos of management that is based on high levels of trust which recognises that managing people fairly and effectively is crucial to skilled work and high performance • Choice flexibility and control over working hours • Autonomy and control over the pace of work and the working environment • Statutory minimum standards to protect the most vulnerable workers against exploitation • Voice for workers in the critical employer decisions that affect their futures. *(Thus, 'good work' is seemingly not simply a matter of removal of aspects that ostensibly represent 'bad work'; rather it is something of a pantechnicon variable that includes contains a wide range of factors and influences concerning society and employers – beyond recognition of the importance of job satisfaction, the possible role of individual perceptions of what constitutes 'good work' is not covered).*

TABLE 1. HEALTH EFFECTS OF WORK vs UNEMPLOYMENT

Authors	Key features (*Additional reviewers' comments in italics*)
Table 1a: Work *continued*	
(Coats & Max 2005) Discussion paper	**Healthy work: productive workplaces. Why the UK needs more 'good jobs'** (The Work Foundation and The London Health Commission) Brings together thinking on the relationship between health, work and productivity, based on research and development work on the nature of good jobs, productivity and the role of work in improving health. Work is better than worklessness and a good job is better than a bad job. There is an important social gradient in health. Bad jobs are characterised by: poor pay; insecure employment; monotonous and repetitive work; lack of autonomy, control and task discretion; imbalance between workers' efforts and the rewards they receive; absence of procedural justice in the workplace. These factors contribute to the social gradient in health and are also linked to productivity. There is an economic, business and a public health case for higher quality employment. Employers and businesses have important and distinctive roles in promoting health and well-being and in tackling health inequalities. These are also now receiving higher political priority. Issues a challenge to government, employers, and unions to rethink their whole approach to management, job design, skills development, and skills utilisation. Also calls for a more sophisticated public debate about the linkages between work and health.

TABLE1. HEALTH EFFECTS OF WORK vs UNEMPLOYMENT

Authors | Key features (*Additional reviewers' comments in italics*)

Table 1b: Unemployment *continued*

(Jahoda 1982)
Monograph

Employment and unemployment: a social-psychological analysis

(*Early, much-quoted classic. Historical and theoretical analysis of the psychology of work and unemployment, but very little actual evidence.*) Historically, unemployment was dominated by inadequate standard of living and poverty, and Jahoda added the psychological impact of being without a job. She started from the premise that 'work is man's strongest tie to reality' (Freud); therefore unemployment leads to 'loosened grip on reality' (Jahoda). The manifest and generally taken for granted consequence of employment is financial remuneration, allowing the individual to earn a living. But Jahoda listed five further benefits of employment that meet corresponding human needs: 1. work imposes a time structure on the working day; 2. work provides social contacts and relationships beyond the family; 3. work involves the worker in collective efforts greater than he or she could achieve alone; 4. work assigns social status and an important part of personal identity; and 5. work enforces regular purposeful or productive activity. In these ways, Jahoda argued that work promotes psychological health, even when it is bad (though she also argued for the need to improve its quality and to humanise work); her central contention was that unemployment leads to loss of these benefits and fails to meet these needs and so is harmful to psychological health.

(*Ezzy 1993 (see below) criticised Jahoda's model as assuming that work is always good and unemployment always bad, and for assuming direct psychological consequences on mental health, while failing to allow for social aspects or the individual's own experience and interpretation of unemployment*).

TABLE1. HEALTH EFFECTS OF WORK vs UNEMPLOYMENT

Authors	Key features (*Additional reviewers' comments in italics*)

Table 1b: Unemployment *continued*

(Hakim 1982)
Narrative review

The social consequences of high unemployment

(*Societal perspective with particular emphasis on UK studies.*) The most immediate consequences of unemployment are those experienced by the unemployed and their families, and this burden is highly concentrated in particular groups, mainly unskilled and manual workers (*who commonly have multiple disadvantages* – (Ashworth *et al.* 2001; Berthoud 2003; Dean 2003; Grewal *et al.* 2004)). However, protracted high levels of unemployment also carry broader social consequences for society as a whole:

- The great majority of unemployed people suffer a significant drop in income which pushes many into poverty; thus, high unemployment rates increase the numbers in poverty;

- There is substantial evidence that as unemployment rates rise, so does the incidence of morbidity and mortality;

- There is a great deal of evidence that prolonged unemployment is commonly a demoralising and stigmatising experience that affects people's will to work, and self-confidence in seeking and gaining work.

- More generally, self-respect, personal, social and work-related skills are eroded by long periods of unemployment.

- All studies of the impact of unemployment on family relationships show an increase in friction, stress and tension, particularly between spouses, but to a lesser extent between parents and children.

- There is extensive evidence (mainly from the US but also from the UK) that high levels of unemployment contribute to higher levels of crime and delinquency and a rise in the prison population.

- There is more patchy evidence that unemployment may have other social consequences, including homelessness, family stability, children's education, racial tension, and public attitudes.

TABLE1. HEALTH EFFECTS OF WORK vs UNEMPLOYMENT

Authors	Key features (*Additional reviewers' comments in italics*)

Table 1b: Unemployment *continued*

(Brenner & Mooney 1983) Narrative review	**Unemployment and health in the context of economic change** Overview of evidence relating unemployment to health at every level of social science analysis from aggregate, population-based studies (macro level), to organisation-based studies to individual-based studies (micro-level). At the population level, increasing unemployment rates indicate recession, economic instability, and/or structural economic decline. At the individual level, unemployment may indicate social stress or downward social mobility and is a stressful life event. At both levels, there is an inverse relationship between measures of economic growth, socio-economic status and differentials, socio-cultural change, economic instability, unemployment, social stress and work-related stress and various indicators of physical and mental health and mortality. Further research should examine a broader range of severity of both 'unemployment' and of health outcomes, and should identify and measure the effects of conditional and modifying factors.

TABLE 1. HEALTH EFFECTS OF WORK vs UNEMPLOYMENT

Authors	Key features (*Additional reviewers' comments in italics*)

Table 1b: Unemployment *continued*

(Platt 1984) Narrative review	**Unemployment and suicidal behaviour: a review of the literature** Comprehensive review of early literature on suicidal behaviour (deliberate self-harm acts): both suicide (fatal outcome) and parasuicide (non-fatal outcome). Considered cross-sectional individual, cross-sectional aggregate, longitudinal individual and longitudinal aggregate studies and their methodological problems. Much of the data at that time related to males, partly because of difficulty defining the economic activity status of married women/housewives. Cross-sectional individual studies showed significantly more parasuicides and suicides are unemployed than expected from general population studies. Likewise, parasuicide and suicide rates are considerably higher among the unemployed than the employed. Increased duration of unemployment was associated with increased rates of parasuicide. Cross-sectional aggregate studies showed a significant geographical association between unemployment and parasuicide. All but one longitudinal individual studies showed significantly more unemployment, job instability and occupational problems among people who committed suicide. Longitudinal aggregate studies show a significant positive association between unemployment and suicide in the US and some European countries; the negative relationship in the UK during the 1960s and early 70s was considered to be due to a unique decline in suicide rates because of the reduced availability of domestic gas, which was previously the most common method of suicide. In conclusion, this review found strong evidence of an association between unemployment and suicidal behaviour, but the limited longitudinal evidence available at that time did not permit any firm conclusions about the causal nature of this relationship. Left open the question of: 'is it the uncertain nature of unemployment, the behavioural reaction (more drink, more cigarettes) to being without a job, or the fact of relative poverty – – or is it some complex interaction of all three?' Most of the included studies were of unemployment, but this review included a number of studies showing similar findings related to 'related socio-economic factors' and 'problems at work' including financial and job worries, work accidents, occupational discontent, job instability, (early) retirement, disciplinary proceedings and dismissal. There were insufficient studies and data to analyse any of these individually, but eight cohort studies identified some form of financial worries as a 'main precipitant cause' leading to parasuicide. Nevertheless, the overall impression of this review was that unemployment and financial problems are only one and usually not the sole or even major precipitants of parasuicide; over 90% of suicide victims suffer from a major psychiatric illness and the most important single trigger identified by parasuicides is interpersonal conflict (though there was no evidence on whether the psychiatric illness or interpersonal conflict might be secondary to unemployment or financial problems).

TABLE 1. HEALTH EFFECTS OF WORK vs UNEMPLOYMENT

Authors	Key features *(Additional reviewers' comments in italics)*

Table 1b: Unemployment *continued*

(Smith 1985) BMJ mini-series (Smith 1987) Monograph	**Occupationless health: a disaster and a challenge** "Bitterness, shame, emptiness, waste": an introduction to unemployment and health. (Educational series that provided a limited review of the scientific evidence, but had a major impact at the time on raising awareness and promoting interest in issues around unemployment and health and their significance for health care and clinical management).
(Warr 1987) Book (Warr 1994) Theoretical paper	**Work, unemployment and mental health** In Warr's 'vitamin model', jobs and unemployment can be characterised as psychologically 'good' or 'bad' on nine dimensions: opportunity for control, opportunity for skills use, externally generated goals, variety, environmental clarity, availability of money (financial hardship in unemployment arguably having the most important impact on mental health), physical security, opportunity for inter-personal contact, and valued social position. Each of these can influence various dimensions of mental well-being (often in a curvilinear manner) subject to modification by various personal characteristics. Warr's central contention is that mental health requires a sufficient level of these environmental 'vitamins' but in some instances an excess can also be harmful. (*For a more succinct and slightly updated summary of the vitamin model, see Warr 1994*). Warr reviewed evidence that unemployment generally has negative impacts on mental health, including lower affective well-being, self-esteem, life satisfaction and happiness, psychological distress, anxiety and depressed mood. There was limited evidence at that time on competence, autonomy and aspiration. However, these negative effects were not universal: they depend on the quality of the jobs and of unemployment; and a small minority of people (5-10%) were found to have improved mental health after losing their jobs. There was consistent evidence that unemployment has a particularly negative effect on middle-aged men, especially those with dependent families. Unemployed teenagers generally show less immediate impact on mental health, but may be delayed developing full adult citizenship. There was less research at that time into unemployment in women: job loss in married women with family responsibilities generally has less impact on mental health; single women with no family responsibilities may be more comparable to men. Job loss generally has a rapid impact on mental health, which deteriorates for 3-6 months and then plateaus; the long-term unemployed may then adapt to their jobless situation and have more limited psychological resources for re-entering work. (*Ezzy 1993 (see below) considered that Warr's vitamin model was the most sophisticated psychological model of unemployment, which allows for many of the beneficial and harmful mental effects of work and unemployment. However, he criticised it as being very 'situation-focused' and failing to allow for the subjective experience of the unemployed individual and for the social aspects of work and unemployment. That criticism appears to be overstated (see (Warr 1994))*.

TABLE1. HEALTH EFFECTS OF WORK vs UNEMPLOYMENT

Authors	Key features (*Additional reviewers' comments in italics*)

Table 1b: Unemployment *continued*

(Catalano 1991) Systematic review	**The health effects of economic insecurity** Economic insecurity includes recession, rising unemployment rates and job loss (*though economic insecurity was not precisely defined nor operationalised for this review*). There is strong evidence that job loss is a significant risk factor for reporting symptoms of psychological distress (but with caveats about the seriousness of these symptoms, the strength of the effects and the generalisability of the findings). There is strong evidence from community level studies that economic recession leads to increased consultation rates with mental health services. There is strong evidence from aggregate level studies that economic recession is associated with increased suicide and parasuicide rates (though with some conflicting evidence). Concluded that the health effects of economic insecurity are mediated by economic policies, but the evidence is insufficient to estimate the likely impact of policy alternatives.

TABLE 1. HEALTH EFFECTS OF WORK vs UNEMPLOYMENT

Authors | **Key features** *(Additional reviewers' comments in italics)*

Table 1b: Unemployment *continued*

(Ezzy 1993)
Theoretical &
conceptual narrative
review from
sociological
perspective

Unemployment and mental health: a critical review

Simplistic descriptions of work as 'good' and unemployment as 'bad' do not adequately explain the observed effects of unemployment on mental health. The majority of unemployed people experience lowered psychological well-being, but a significant minority show improved mental health. Similarly, re-employment typically restores original levels of mental health, but some re-employed people report lowered mental health. Employment is not unambiguously positive and unemployment is not unambiguously negative. The experience of *(employment and)* unemployment varies considerably depending on age, gender, (education & social background), income, social support, reason for job loss, commitment to employment, satisfaction with previous work, expectation about returning to work and duration of unemployment. However, it is not clear whether these moderator variables reflect variations in the objective circumstances of unemployment or whether they influence the individual's psychosocial response. Some people become adapted to economic inactivity, finding it less unpleasant than employment, particularly if they could only obtain poor quality work in oppressive conditions where the financial rewards are little if any better than social security benefits.

Ezzy provided a critique of various previous psychological models:

1. 'Rehabilitation' approaches: individual-centred, focusing on personal characteristics of the unemployed. These fail to address structural causes of unemployment.

2. 'Stages' models (Eisenberg & Lazarsfeld 1938): these really offer a descriptive framework rather than theoretical understanding or models of processes or mechanisms. They commonly describe three stages that unemployed persons pass through over time, which may be broadly described as: optimistic activity; increasing distress; fatalism and apathy. Stages models generally fail to allow adequately for the variation in individual response to unemployment and may lead to stereotyping. They are also based mainly on evidence from older people and may not apply to unemployed youths. *(These comments on stages models include some from Lakey et al 2001).*

3. Jahoda's functional model *(See Jahoda 1982 above)*.

4. Warr's 'vitamin model' *(See Warr 1987 above)*.

5. Various psychological models that emphasise cognitive processes, personal control, and coping. These develop some important issues but are generally narrow and specific and fail to allow adequately for other factors.

Ezzy then proposed a sociological model according to which the cause of poor mental health in many unemployed people is considered to be rooted in failure to find meaning in life, and the experience of unemployment is a transition phase between the loss of meaning associated with employment and the reintegration associated with either re-employment or adaptation to an alternative lifestyle. (Again, there appears to be some validity in this, but it is a narrow and specifically sociological perspective).

TABLE1. HEALTH EFFECTS OF WORK vs UNEMPLOYMENT

Authors	Key features (*Additional reviewers' comments in italics*)

Table 1b: Unemployment *continued*

Authors	Key features
(Bartley 1994) Conceptual, narrative review	**Unemployment and ill-health: understanding the relationship** Four mechanisms need to be considered: 1. Poverty. There is strong evidence that the (psychological) health effects of unemployment are at least partly mediated through relative (rather than absolute) poverty and financial anxiety (c.f. health inequalities). 2. Stress. Even apart from any financial impact, there is strong evidence that unemployment is stressful, with direct effects on psychological health. Unemployment may also affect physical health via a 'stress' pathway involving physiological changes such as raised cholesterol and lowered immunity. 3. Health related behaviour, including that associated with membership of certain types of sub-culture. 4. The effect that a spell of unemployment has on subsequent employment patterns. (Life course perspective). *(The Abstract also listed social isolation and loss of self-esteem as one of the main mechanisms, but this was not considered further in the paper).*
(Hammarström 1994b) Narrative review	**Health consequences of youth unemployment – review from a gender perspective.** The association between unemployment and ill-health is explained by both selection and exposure; gender may be a confounding factor in both selection and exposure. Young people generally start in a better state of (at least physical) health, and the ill effects of unemployment are therefore likely to be less than in adults. Most studies focus on individuals and psychological ill health and there is relatively little research on somatic health (where the limited evidence is conflicting), societal or familial consequences, or investigating theoretical models. There is consistent evidence of an association between unemployment and minor psychological disorders, which may be greater among young women and among young men with lower education. Unemployed young men show worse health behaviour on most measures (e.g. eating habits, personal hygiene, sleeping habits, physical activities). Increased health care consumption has been documented. Unemployment is a risk factor for increased alcohol consumption, especially among young men, increased smoking and increased use of illicit drugs. The mortality rate is significantly higher in both young men and young women, mainly due to accidents and suicide. Social consequences include alienation, lack of financial resources, criminality and future exclusion from the labour market. Social support, high employment rates, negative attitudes to work and high possibility of control act as mediating factors with a protective effect on health. There is a need for more qualitative research based on theoretical models, to understand causal mechanisms that determine health inequalities, in which unemployment is one important factor.

TABLE1. HEALTH EFFECTS OF WORK vs UNEMPLOYMENT

Authors	Key features (*Additional reviewers' comments in italics*)

Table 1b: Unemployment *continued*

Authors	Key features
(Jin *et al.* 1995) Systematic review	**The impact of unemployment on health** (*Included 46 original studies*) Epidemiological evidence shows a strong, positive association between unemployment and many adverse health outcomes. Most aggregate-level studies report a positive association between national unemployment rates and rates of overall mortality, cardiovascular disease mortality, and suicide. Large, census-based, longitudinal, cohort studies show higher rates of overall mortality (5 studies), cardiovascular mortality (4 studies) and suicide (3 studies) among unemployed men and women than among either employed people or the general population. Four individual-level studies show elevated levels of intermediate cardiovascular outcomes such as blood pressure or serum cholesterol. Two studies found increased cerebrovascular disease mortality. One study found higher mortality rates among the wives of unemployed men. Nine individual-level, longitudinal studies consistently found an association between unemployment and general health and also mental health problems. Workers laid off because of factory closure (4 studies) report more symptoms and illnesses than employed people. Unemployed people may be more likely to visit physicians, take medications, and be admitted to general hospitals. There is conflicting evidence on a possible association between unemployment and admission rates to psychiatric hospitals and alcohol consumption, which may be complicated by other institutional and environmental factors. The authors then assessed this evidence on the basis of epidemiological criteria for causation, and concluded that whether unemployment causes these adverse outcomes is less straightforward, because there are likely many mediating and confounding factors, which may be social, economic or clinical. Health selection may be a particular problem in the association between psychiatric illness and unemployment. Many previous authors have suggested possible causal mechanisms, but further research is needed to test these hypotheses.
(Banks 1995) Narrative review	**Psychological effects of prolonged unemployment: relevance to models of work re-entry following injury** Integrates research and theory into the psychological effects of unemployment with theoretical models of work disability. Unemployed people have lower levels of affective well-being, higher levels of psychological distress and lower self-esteem; longitudinal studies show that unemployment causes these effects while re-employment improves them. These effects are mediated by personal factors such as age, gender, ethnic group, social class, social relationships, duration of unemployment, employment commitment, local unemployment rate and personal vulnerability. Environmental factors (as in *Warr's* 'vitamin model') can also influence mental health and interact with the direct effects of unemployment. The principal argument of this paper is that these psychological findings help to understand and explain the dynamics of individual adjustment and return to work following injury and illness. Particularly relevant are distancing from the labour market, loss of control and motivation, externally generated goals and task variety, loss of income, the impact of the social security system, financial and material deprivation, and loss of social status. This has implications for early intervention, rehabilitation, and for flexible and structured return to work and retraining programmes.

TABLE1. HEALTH EFFECTS OF WORK vs UNEMPLOYMENT

Authors	Key features (*Additional reviewers' comments in italics*)

Table 1b: Unemployment *continued*

(Shortt 1996) Conceptual, narrative review	**Is unemployment pathogenic? A review of current concepts with lessons for policy planners** Emphasised that policy makers and planners should have some knowledge of 'the complex nexus between ill-health and unemployment' and that the relationship should not be viewed as unproblematic. (*After detailed methodological criticism (e.g. of Brenner's studies) and discussion, Shortt nevertheless concluded that*): 1. 'It seems clear – that for selected groups, for selected causes, in several nations, unemployment accounts for at least some increase in mortality rate.' 2. 'It is reasonably well documented – that unemployment, at least among males, has an adverse effect on physical health, particularly with reference to the cardiovascular system.' 3. 'Unemployment – clearly has an adverse effect on – mental health - – though the relationship - - (is) complex and contingent upon prevailing social or medical resources.' 4. ' – most (*unemployed*) women suffer some adverse mental and physical effects.' 5. 'Like their elders– it is clear that adolescents and young adults experience both physical and especially mental ill-health as a result for unemployment.' 6. 'The existing literature – while deficient in many areas, is strongly suggestive of a pathological impact for employment on the children and families of the unemployed.' 7. 'Unemployment engenders increased utilization of medical services, particularly primary care, in those countries where access is not dependent on personal finances.' There was limited evidence on the causal mechanisms between unemployment and these adverse health effects. He hypothesised that the effects might be related to social gradients in health rather than poverty *per se* (Marmot 2005). But he also pointed out this might be modified by individual circumstances or social context. Policy makers must face the reality that 'unemployment deserves to be considered a significant social cause of ill-health' and economic policy planning should take this into account. However, it must avoid certain pitfalls: • Policy must not medicalise unemployment, which is ultimately a social process. • The long-term goal is not to 'normalise' unemployment by making it more 'comfortable' but to reduce its incidence. • In identifying the adverse health effects of unemployment, it must not be forgotten that employment also carries hazards and risks to health, and that the quality of work may be as important for health status as unemployment.

TABLE1. HEALTH EFFECTS OF WORK vs UNEMPLOYMENT

Authors	Key features (*Additional reviewers' comments in italics*)

Table 1b: Unemployment *continued*

Authors	Key features (*Additional reviewers' comments in italics*)
(Weber & Lehnert 1997) Quasi-systematic review	**Unemployment and cardiovascular diseases: a causal relationship?** This extensive literature analysis starts from the position that although negative effects on social life and psychological variables resulting from unemployment are generally accepted, a possible causative relationship between job loss and somatic illness remains controversial. The basic question addressed is: does somebody become ill because he is unemployed (causality hypothesis) or does he become unemployed because he is ill (selection hypothesis). 10 person-related studies that focused on whether unemployment can influence cardiovascular morbidity were included (7 of which were longitudinal studies with 2 to 8 years follow-up). The methodological aspects of these studies were considered less than ideal. In some cases statistically significant associations were found between unemployment and the increase in cholesterol levels or systolic/diastolic blood pressure, but the clinical relevance of such slight changes is questionable. Authors concluded (*at that time*) that consideration of unemployment as an independent, social, cardiovascular risk factor was unwarranted. An increase in the prevalence rates of coronary heart disease or arterial hypertension causally linked in some studies with unemployment is scientifically questionable due to severe methodological shortcomings.
(Lynge 1997) Systematic review	**Unemployment and cancer** Unemployed men have an excess cancer mortality of close to 25% (mainly from lung cancer) compared with all men in the labour force. The available data suggest that applies whether the unemployment rate is 1% or 10%. The risk persists long after the start of unemployment and does not disappear after controlling for social class, smoking, alcohol intake, and previous sick leave.
(Mathers & Schofield 1998) Brief narrative review	**The health consequences of unemployment** The relationship between unemployment and health is complex: ill-health also causes worklessness and confounding factors include socio-economic status and lifestyle. Nevertheless, longitudinal studies with a range of designs provide reasonably good evidence that unemployment *per se* is detrimental to health and impacts on a number of health outcomes – increasing mortality rates (those unemployed who had a pre-existing illness or disability had mortality rates over three times higher than average. Those who were unemployed but not ill showed a 37% excess mortality over the following 10 years), causing physical (self-reported 'poor health', limiting long-standing illness, objective cardiovascular disease, lung cancer) and mental ill-health and increased use of health services. (*Systematic literature search with narrative review of key studies and a particular focus on Australian evidence*).

TABLE1. HEALTH EFFECTS OF WORK vs UNEMPLOYMENT

Authors	Key features (*Additional reviewers' comments in italics*)
Table 1b: Unemployment *continued*	
(Morrell *et al.* 1998) Brief narrative review	**Unemployment and young people's health** There is strong evidence of an association between unemployment and ill-health in young people aged 15-24 years. Aggregate data show a strong association between youth unemployment and suicide. Youth unemployment is also associated with psychological symptoms such as depression and loss of confidence (though these are less severe and in some cases different from in unemployed adults). (*Many of the studies of youth unemployment and mental health are from Australia.*) Effects on physical health have been less extensively studies, but there is limited evidence on raised blood pressure. There is inconsistent evidence of an association between unemployment and lifestyle risk factors such as increased cannabis, tobacco, and alcohol consumption. (*Systematic literature search with narrative review of key studies and a particular focus on Australian evidence*).
(Björklund & Eriksson 1998) Narrative review	**Unemployment and mental health: evidence from research in the Nordic countries** Detailed discussion of the methodological problems of establishing cause and effect (*by economists*). Cross-sectional studies show that unemployed people have worse mental health than do others. Most longitudinal studies suggest that unemployment is associated with deteriorating mental health, though it is somewhat unclear how long such an effect lasts. There was insufficient evidence to reach any conclusion on the relative importance of heterogeneity or duration dependency on exit rates from unemployment. Unemployment benefits and labour-market policy affect the pattern of exit rates out of unemployment. (*Similar findings and conclusions to other reviews of the international evidence*).
(Cohen 1999) Narrative review	**Social status and susceptibility to respiratory infections** Lower social status (including unemployment, perceived and observed social status) in human adults and children and other primates is associated with increased risk of respiratory infections, thought to be due to a combination of increased exposure to infectious agents and decreased host resistance to infection.

TABLE1. HEALTH EFFECTS OF WORK vs UNEMPLOYMENT

Authors	Key features (*Additional reviewers' comments in italics*)

Table 1b: Unemployment *continued*

(Nordenmark & Strandh 1999) Theoretical paper and empirical analysis	**Towards a sociological understanding of mental well-being among the unemployed: the role of economic and psychological factors** Classic psychological research has focused on the psychological importance of work and the negative impact of unemployment, but the effects are not the same for everyone. The adverse effects on mental well-being may be mediated by the individual's economic situation, gender, social class, age, marital status, duration of unemployment, previous history of (un)employment, ethnicity, and work involvement; different individuals may respond differently. This paper aimed to develop a theoretically and empirically founded sociological model with good predictive powers for explaining both the differences in mental well-being among the unemployed and the changes in individual mental well-being during unemployment and upon re-entering employment. This model was tested empirically in a longitudinal study of 3,500 Swedes. The model starts with a socially constructed individual identity with social roles and social goals, which are heavily influenced by the individual's position in society. Employment is then one important resource for meeting the individual's socially defined needs in two ways: a) it is the main provider of economic resources that enable the individual to participate in society; b) it meets important psychological needs in a society where employment is the norm (see Jahoda 1982; Warr 1987). Unemployment may fail to meet these needs; alternatively, needs may be met in other ways or individuals can redefine their social roles and goals in which work is less important. Mental well-being depends on the balance between needs and resources to meet them. The final model integrates both the economic need for employment and psychosocial needs for social roles and social goals, and shows that their combined effect is central to mental well-being. (*A sociological analysis of (un) employment*).

TABLE1. HEALTH EFFECTS OF WORK vs UNEMPLOYMENT

Authors	Key features (*Additional reviewers' comments in italics*)

Table 1b: Unemployment *continued*

(Murphy & Athanasou 1999) Systematic review and meta-analysis	**The effects of unemployment on mental health** Included 16 longitudinal studies from 1986-96 on the mental effects of unemployment. Many studies included both men and women; most were of general mental well-being, psychological distress and/or depressive symptoms (e.g. General Health Questionnaire GHQ)); 3 studies were in UK. 14 out of 16 studies showed that job loss had, on average, a negative effect on mental health; 5 of these studies on 616 subjects showed a weighted average effect size of d = 0.36. Seven studies on 1509 subjects showed that re-employment had, on average, a positive effect on mental health with a weighted average effect size of d = 0.54. Both of these effect sizes (where d is the difference in group means on standard psychometric tests divided by standard deviation) are of 'practical significance'. The major methodological question was how far these effects might be explained by selection bias - i.e. people with poorer (*or deteriorating*) mental health are more likely to lose their jobs, while people with better (*or improving*) mental health are more likely to be re-employed - though at least 4 of the studies attempted to allow for this. Other studies raised questions about the context of unemployment (e.g. country or local unemployment rate) and about the quality of new jobs.

TABLE1. HEALTH EFFECTS OF WORK vs UNEMPLOYMENT

Authors	Key features *(Additional reviewers' comments in italics)*

Table 1b: Unemployment *continued*

(Lakey 2001) Policy Studies Institute Monograph	**Youth unemployment, labour market programmes and health** Patterns of youth employment vary: it may be permanent or temporary, full-time or part-time, secure or insecure, well or poorly paid, and offer good or poor experience and prospects for career progress; even if not currently in paid work, routes to future economic activity include further education, training schemes, and voluntary work. Patterns of youth unemployment also vary: some young people experience a single or multiple periods of short- or long-term unemployment; some have experienced one or more jobs and job losses, but others have never worked at all. Poor health may lead to unemployment and unemployment may adversely affect health, but there may also be an interaction between them to produce vicious circles of unemployment and health deterioration (perhaps especially with mental health problems). Most of the research on the relationship between unemployment and health, particularly among young people, has focused on mental health. Unemployed young people experience more health problems than those who are employed: lower levels of general health (as shown on symptom checklist scores and on self-rated health, disability, long-standing illness, limiting long-standing illness, medical consultation, and hospital admission rates*); more anxiety and depression; higher rates of smoking; and higher suicide rates. Young people with health problems have less success in finding jobs and are more likely to lose or leave their jobs. Longitudinal studies show that unemployment is also detrimental to the health of young people. Financial stress, material and experiential deprivation contribute to the detrimental health effects. Overall, the health effects of unemployment on young people appear to be less than on adults (perhaps because they start generally healthier, *and have different social and financial commitments*), but young people from disadvantaged backgrounds, those with lower levels of education, or those who lack social support may be particularly vulnerable. There is a need for further research on sub-groups, including gender. Labour market and health interventions have the potential to reverse the downward spiral of poor health and unemployment, but more research is required to identify their specific health effects. (* *Note that these measures of general health are all highly correlated with psychological well-being and should be distinguished from physical health*) (*Systematic search and narrative review*).

TABLE 1. HEALTH EFFECTS OF WORK vs UNEMPLOYMENT

Authors	Key features (*Additional reviewers' comments in italics*)

Table 1b: Unemployment *continued*

(Brenner 2002) Report to the European Commission

(*Un*)-Employment and public health

The major portion of this 800 page Report consists of economic time series analyses (quantitative historical analyses of data over time – longitudinal aggregate studies) of the relationship between employment and unemployment rates and mortality patterns in 15 western European countries and the United States since the Second World War. These showed that mortality rates are related to real GDP per capita, unemployment rates, and the interaction between real GDP per capita and the unemployment rate. However, all countries studied showed a relationship between unemployment and mortality, even after holding constant the impact of real GDP per capita and the interaction between real GDP per capita and the unemployment rate. Unemployment rates are historically and contemporaneously related to elevated mortality rates: the higher the unemployment rate in a given year, the higher the mortality rate over the following 10–15. Suicide increases within a year of job loss, and cardiovascular mortality accelerates after two or three years and continues for the next 10–15 years. Thus, increasing employment rates in the direction of a "full employment" economy is a fundamental source of decreased mortality, and increased rates of unemployment are related to heightened mortality rates and thus decreased life expectancy. Brenner concludes that changes in the national economy, especially employment and unemployment rates (e.g. 'cyclic' unemployment related to national and international recessions) but also economic growth, structural and technological factors are the major influence on mortality patterns and life expectancy. Put another way, the economic and social status of families is of great potential importance in their health and life expectation. Brenner suggests that this is largely a matter of material standards of living: 1) investment in improved working conditions, 2) purchase of items that will increase the comfort, mobility and functioning of elderly, frail and disabled populations, 3) improved nutrition, 4) major investments in scientific development and public education, 5) increase of sophistication of, and population access to, health care technology, and 6) increased financial security and reduction of poverty through social welfare/security systems. (*Findings for UK were broadly consistent with other countries*). (*Highly technical economic modelling, with extensive debate about the strengths and weaknesses of the methodology* (e.g. *see* (Platt 1984; Wagstaff 1985; Ezzy 1993; Shortt 1996)). *The major strengths are that it provides evidence at a societal level and on the effect of social policy. The major weaknesses are that it only considers the impact of unemployment (and not sickness or interactions between work and health), focuses almost exclusively on mortality, and fails to provide any evidence about the impact on individual physical or mental health*).

TABLE1. HEALTH EFFECTS OF WORK vs UNEMPLOYMENT

Authors	Key features (*Additional reviewers' comments in italics*)

Table 1b: Unemployment *continued*

(Brenner 2002) Report to the European Commission	*(Un)*-**Employment and public health** *continued* Brenner also reviews labour market policy literature and considers implications of his whole study for active labour market policy (*which is outwith the scope of the present review*). Various appendices provide systematic reviews of recent scientific literature on a) employment, socio-economic factors and health (80 papers from 2000-2002; b) the causal relationship of unemployment and health status (63 papers from 1994-2001 – *proper reference list missing*); and c) the influence of health services on employability (89 papers from 1994-2001 – *proper reference list missing*). (*These reviews include odd selections of papers: the large majority are cross-sectional or uncontrolled cohort studies, many are highly condition or country specific, many are of little relevance to the present review*). Brenner concludes that there is 'massive evidence' in the scientific literature that, for many causes of acute and chronic illness and disability, appropriate health care could materially improve the health status of the unemployed, and also substantially improve their employability. (*That is completely unfounded on the evidence presented. Brenner's conclusions regarding the likely impact of health care on occupational outcomes - especially for common health problems - appear questionable*).
(Saunders 2002b), (Saunders & Taylor 2002). Discussion papers	**The direct and indirect effects of unemployment on poverty and income inequality** (Australian Social Policy Research Unit) In a world in which people's sense of identity and the material prosperity they are able to enjoy are inextricably linked to their status as workers, consumers and citizens, unemployment can be profoundly debilitating. Unemployment has high economic costs to individuals, their families, their (local) communities and the state; but unemployment does not affect all groups similarly, so its effects are economically unequal and potentially socially divisive. The relationships between unemployment, poverty, inequality and social exclusion are complex (partly because of questions about how these issues are conceptualised and measured), but there are sound reasons to expect high levels of unemployment to be associated with increased poverty, greater inequality and more exclusion. There is strong evidence that unemployment leads to reduced income, increases the risk of poverty and contributes to inequality, and that it also gives rise to a series of debilitating social effects, including social exclusion, impacts on family life and the cohesion of families with an unemployed member, and on the nature of local communities affected by widespread and systemic unemployment, including the consequent increase in crime rates that often accompanies geographical concentrations of unemployment. These social effects interact with and reinforce each other, making it harder to reverse the pattern of events that originally gave rise to them. Thus, unemployment adversely affects morale and health, making the prospect of re-employment less likely, whilst simultaneously leading to attitudes that reinforce detachment from the world of work. The more that unemployment becomes a structural feature of society, the harder it is for individuals to escape its effects. Concludes that 'unemployment is a bad thing. It is bad for the economy and for society, for unemployed people themselves, for their families and for the communities in which they live.'

TABLE 1. HEALTH EFFECTS OF WORK vs UNEMPLOYMENT

Authors	Key features (*Additional reviewers' comments in italics*)

Table 1b: Unemployment *continued*

(McLean *et al.* 2005) Evidence review and report

Worklessness and health: what do we know about the causal relationship?
(Health Development Agency, UK)

This report aimed to provide a synopsis of the evidence about the causal relationship between worklessness and ill health, which was intended to inform policy and decision makers, organisations with an interest and remit for employment and health, and employers in the widest sense. Using the Health Development Agency evidence review methodology, it was an overview of review level evidence. Because of the methodological criteria used, the report focused on the literature about unemployment and only included 12 reviews published between 1990 and 2003. The report's main conclusions were:

- The evidence outlined in this review shows a relationship between unemployment and poor health, although causation is not proven.

- There would seem to be a strong relationship between psychiatric morbidity and unemployment.

- Much of the evidence from both original studies and reviews (*i.e. those included in this report*) deals with the concept of unemployment, and not worklessness in its broadest sense.

- There is a need for an increased sophistication in understanding the health and work agenda within the context of health inequalities, especially the geographical dimension. Improvements in the nation's health may not by itself have a significant impact on health inequalities.

- There is a strong association between deprived areas, poor health, poverty and worklessness although the exact relationship is not clear.

- (*In summary*), the evidence suggests a relationship between unemployment and poor health, with a strong association between unemployment and poor mental health. The relationship though is complex and unclear, in part confounded by other variables such as educational attainment, the environment, and economic circumstance.

TABLE 1. HEALTH EFFECTS OF WORK vs UNEMPLOYMENT

Authors

Key features (*Additional reviewers' comments in italics*)

Table 1b: Unemployment *continued*

(Ritchie *et al.* 2005)
DWP Research Report

Understanding workless people and communities: a literature review
(Department for Work and Pensions)

Review of evidence relating to the psychological and social influences on workless people in deprived areas. Although employment rates have been increasing since 1992, there are geographical areas of persistent high unemployment and a hard core of long-term unemployed. Worklessness was defined as detachment from the formal labour market: economic inactivity may include those who have never worked, the short or long-term unemployed (and claiming unemployment benefits), sick or disabled (whether or not claiming various disability and incapacity benefits), those who chose not to work (e.g. those with family responsibilities, carers or approaching retirement), and those who are actually working but solely in voluntary work or the informal economy. Traditional psychological models of job loss and mental health (e.g. Jahoda 1982; Warr 1987) may not hold true in other forms of worklessness or where 'not being in employment' is the norm. To understand the impact of worklessness, it is necessary to look at individual behaviour in the broader context of the communities and areas in which these people live. Groups at higher risk include lone parents, minority ethnic groups, disabled people, carers, older workers, workers in the informal economy, offenders and ex-offenders – many of whom suffer multiple disadvantages and barriers to participating in the labour market. Although it is clear that worklessness generally has a negative impact on well-being, there is a lack of evidence on how persistent high local unemployment rates and poor employment prospects influence the impact of worklessness on individuals. There is limited evidence on people who have been out of work more than 3 years or who have never worked. There is conflicting evidence about the existence of a 'culture of worklessness' – lowered incentives to work where peers are also unemployed and the informal economy has a strong pull factor, a view that joblessness is unproblematic within a context of lowered aspirations, and short-term horizons.

- The causes of persistent worklessness transcend personal and psychological characteristics.
- Persistent worklessness in the face of labour market buoyancy opportunities suggests that the barriers and constraints to (return to) work are likely to be complex, multifaceted, deep-rooted and individually varied.
- Workless people often have problematic experiences of (unsatisfactory) work.

TABLE 1. HEALTH EFFECTS OF WORK vs UNEMPLOYMENT

Authors	Key features (*Additional reviewers' comments in italics*)

Table 1c: Older workers (> approx. 50 years)

(Tuomi *et al.* 1997) Research summary	**Summary of Finnish project to promote the health and work ability of ageing workers**

(Not strictly a review, rather a summary of a large Finnish longitudinal study of ageing workers between 1981 and 1992 which produced a series of papers). 6,259 workers age 44–58 at baseline, in 40 occupations, were followed-up; some had retired by the end of the study. After 11 years of ageing the workers felt their work had become heavier both mentally and physically. Ageing was accompanied by the appearance of various diseases (especially musculoskeletal and cardiovascular) and symptoms, but the subjects perceived their health as improved: 42% of those with a diagnosed disease in 1992 perceived their health as good compared with only 11% in 1981. Work ability was measured by the work ability index, which depicts the ability to work from a positive point of view of health and mental resources, and also relates the ability to work to the demands of the work and to disease. The work ability of the subjects deteriorated before retirement age: work does not seem to prevent a decline in work ability with age. Work ability decreased most in those with physical jobs. Some workers improved in work ability – associated with improvement in supervisor attitude, decreased repetitive actions and increased leisure-time exercise. There was a shift from physical jobs to mental jobs among those who remained employed during follow-up. There are positive changes related to ageing in workers: improved perceived health and increased interest in exercise. Muscular demands of work must not exceed physical capacity of ageing workers, and job content and social support should be developed. The impact of work on functional capacity and symptoms of workers may begin even earlier than age 45. (*Unfortunately there was no specific comparison of the health of those who continued working and those who retired*).

TABLE1. HEALTH EFFECTS OF WORK vs UNEMPLOYMENT

Authors	Key features (*Additional reviewers' comments in italics*)

Table 1c: Older workers (> approx. 50years) *continued*

Authors	Key features
(Hansson *et al.* 1997) Narrative review	**Successful ageing at work: annual review, 1992–1996: the older worker and transitions to retirement** Defines the older worker as >40 because it presents a physiological milestone (*though no evidence presented for this statement*). General conclusions: • Older employees need to be viewed as individuals. Job performance and well-being among older workers appear to reflect the fit between one's changing abilities and the demands of the job (though the individual's experience and coping resources can compensate in part). There is a need to assess on an individual basis the consequences of ageing for a continued person-environment fit. Both employer and employee need to assume responsibility for their part in the process (e.g. positive accommodation v keeping fit) • The transition to retirement is seen as occurring in many forms, reflecting a diversity in person and environment variables: 'blurred' retirements; uncertain starts; re-entries; unemployment turning into retirement • The human factors and safety literature suggests that the traditional workplace designed for the average 20 to 40 year old will need to be redesigned as the workforce ages.
(Wegman 1999) Narrative review	**Older workers** (*A review of a range of issues surrounding older workers, with the focus on capabilities*). Popular misconceptions about competence, knowledge, and work capacity play a large role in determining whether older workers remain employed. There is no direct relationship between ageing and decline in occupational capacities. The factors which make advancing age into a handicap are mostly connected with working conditions mismatched with physical capabilities of human beings, and work organization which deny employees any possibility of contributing to the development of their jobs.

TABLE 1. HEALTH EFFECTS OF WORK vs UNEMPLOYMENT

Authors	Key features (*Additional reviewers' comments in italics*)

Table 1c: Older workers (> approx. 50 years) *continued*

Authors	Key features
(Shephard 1999) Narrative review	**Age and physical work capacity** Ageing is associated with a progressive decrement in various components of physical work capacity, including aerobic power and capacity, muscular strength and endurance, and the tolerance of thermal stress. Part of the functional loss can be countered by regular physical activity, control of body mass, and avoidance of cigarette smoking. The various age-related changes are of concern to the occupational physician, because of the rising average age of the labour force. In theory, an over-taxing of the heart and skeletal muscles might be thought to lead to a decrease of productivity, manifestations of worker fatigue such as absenteeism, accidents, and industrial disputes, and an increased susceptibility to musculoskeletal injuries, heart attacks, and strokes. However, in practice, the productivity, health, and safety of the older worker pose relatively few problems. It is stressed that in general there is no longer any need to push workers to their physical limits because of automation-related changes in modern methods of production. Similarly, it is rarely necessary to force the retirement of those who have worked conscientiously for many years, but now find difficulty in conforming to physical – there is an enormous quantity and variety of tasks with only limited physical input.
(Kilbom 1999) Narrative review	**Evidence-based programs for the prevention of early exit from work** Ageing of the population and lowered average age of retirement imply greatly increased public costs for pensions and health care in western societies. Prolongation of working life is necessary to counteract large budget deficits, and most western countries are now in the process of changing public retirement benefits. Researchers acknowledge the importance of work for well-being and health, but also recognise that some working conditions can be detrimental. Any prolongation of working life must be accomplished without threatening the well-being of elderly persons, and therefore working life needs changes that accommodate the capacity and demands of an ageing work force are needed. No scientific intervention studies have as yet demonstrated that early exits from working life can be prevented while work ability, health, productivity, and a high quality of life are maintained. However, several studies on return-to-work after prolonged sick leave, re-entry to work after lay-offs, risk factors for early retirement, risk/health factors for maintained work ability, and case studies provide indirect support for the feasibility of preventing early exits. Information on conditions beneficial for maintaining ageing workers in the workforce is rapidly accumulating.

TABLE1. HEALTH EFFECTS OF WORK vs UNEMPLOYMENT

Authors	Key features *(Additional reviewers' comments in italics)*

Table 1c: Older workers (> approx. 50years) *continued*

(Scales & Scase 2000)
Narrative review

Fit and fifty?

(A comprehensive review of social, occupational, and economic trends among UK adults in their 50s.) It was found that people in their 50s are now comparable in their attitudes, activities and behaviours to people in their 30s and 40s; they don't view themselves as 'older' and it is only in their 60s that they change.

- Twice as many men in their 50s compared with those in their 30s report their health is poor: for women the pattern is less pronounced
- GP visits are not significantly higher for most in their 50s, although they are higher for those living alone and for those who are economically inactive
- Men in their 50s are more likely to experience longer periods of economic activity and ill health but there are occupational differences
- Unemployment leads to low motivation, depression and general disengagement from active social participation
- Moving out of employment reduces stress and improves health levels for men (but not women) in managerial and professional occupations but increases stress and is associated with deteriorating health for those in manual unskilled occupations.
- Satisfaction with life overall runs at 86% (working), 80% (inactive with a pension) and 48% (inactive without a pension)
- For men and women in managerial and professional occupations, and expectation of early retirement has become entrenched and will be difficult to change. Around 25% of those in white collar jobs and about 40% of those in blue collar jobs would give up working if they could afford to do so, although 65% overall would not like to give up working entirely
- For many people in their 50s from professional occupations, early withdrawal from the labour market is a choice based upon access to a occupational or private pension income. However, for manual workers it is more likely to be early retirement on grounds of ill health
- 50% of non-working professionals report that they are financially comfortable compared to <20% of former blue collar workers, 60% of whom are finding things quite or very difficult compared to 30% of inactive professionals.

TABLE1. HEALTH EFFECTS OF WORK vs UNEMPLOYMENT

Authors	Key features (Additional reviewers' comments in italics)

Table 1c: Older workers (> approx. 50years) continued

Authors	Key features
(Ilmarinen 2001) Narrative review	**Ageing workers** The definition of an ageing worker is generally based on the period when major changes occur in relevant work related functions during the course of work life. Functional capacities, mainly physical, show a declining trend after the age of 30 years, and the trend can become critical after the next 15–20 years if the physical demands of work are not reduced. On the other hand, workers' perceptions of their ability to work indicate that some of them reach their peak before the age of 50 years, and by age 55 years about 15–25% report that they have a poor ability to work, mainly those workers in physically demanding jobs but also those in some mentally demanding positions. Therefore, the ages of 45 or 50 years have often been used as the base criterion for the term "ageing worker". The need for early action has been emphasised by the low participation rates of European Union workers who are aged 55-59 years (60%) and those 60-64 years (20%) or older and by the early exit of this age group from work life all over the world. *(Models are presented dealing with means of keeping ageing workers in work – all stakeholders (individual, employer, society) working together to promote work ability; whilst no data are presented on health effects, the underlying assumption is that maintenance of (suitable) work is desirable in older age).*
(Benjamin & Wilson 2005) Narrative review	**Facts and misconceptions about age, health status and employability** (Age Partnership Group, UK) The report considers some of the myths about older workers, and provides evidence and arguments that aim to dispel inaccurate perceptions about older adults. Numerous myths about the relationships between age, health and work are explored and discussed. Older workers do not necessarily have less physical strength and endurance – many jobs are not physically demanding, and physical demands can be minimised through changes in work design. Older workers often show lower levels of short-term sickness absence than younger workers, but some older workers may show more long-term absence – however, chronic diseases that lead to long-term absence are open to workplace interventions. Older workers do not have more accidents at work. It is concluded that older adults are vastly different from each other, and no stereotype is likely to reflect most older workers. There is no health and safety justification to exclude older workers from the workforce, particularly given health and safety legislation requiring employers to minimise the health and safety risks to all employees. *(Apparently aimed at employers in the context of age discrimination in employment legislation to be introduced in 2006).*

TABLE 2. HEALTH IMPACTS OF EMPLOYMENT, RE-EMPLOYMENT AND RETIREMENT

Table 2a: School leavers and young adults (Age < approx. 25 years)

Study	Population/setting	Follow-up	Health measures	Key findings on re-employment (*Additional reviewers' comments in italics*)
(Banks & Jackson 1982, (Jackson *et al.* 1983) England	2 cohorts a) 780 & b) 647 school leavers (male and female)	a) 7, 15 & 30 months b) at school, 8 & 24 months	General Health Questionnaire (GHQ)	School leavers moving into employment, Youth Opportunity Programmes, or further education improved their GHQ scores, while those who became unemployed deteriorated. On moving from unemployment to employment, increased psychological distress fell to levels comparable to those employed throughout Employment commitment amplified or reduced the impact of unemployment on psychological distress.
(Patton & Noller 1984) Australia	57 males & 56 female school leavers	Before leaving school & 5 months	Self-esteem, locus of control and depression	Those moving into employment or returning to school showed improvements in self-esteem, locus of esteem and depression scores. Those leaving school and becoming unemployed showed a greater magnitude of deterioration on all measures.
(Layton 1986a) England	186 male school leavers	Before leaving school & 6 months	GHQ	Baseline measures showed no difference in GHQ scores between those who subsequently obtained work or became unemployed. Youths who were gainfully employed (or who entered further education) showed improvement in mental well-being as a consequence of finding a job.

TABLE 2. HEALTH IMPACTS OF EMPLOYMENT, RE-EMPLOYMENT AND RETIREMENT

Table 2a: School leavers and young adults (Age < approx. 25 years) continued

Study	Population/setting	Follow-up	Health measures	Key findings on re-employment (*Additional reviewers' comments in italics*)
(Donovan *et al.* 1986) Australia unemployed	131 16-year old school leavers	Before leaving school & 6-12 months	GHQ Leeds scale, Life satisfaction, self-esteem.	School leavers who moved into employment showed significant improvements on all measures, while those who became unemployed showed significant deterioration. Those on government training schemes showed health effects that were generally closer to those of employment.
(Feather & O'Brien 1986) (O'Brien & Feather 1990) Australia	3,458 school leavers	Before leaving school, 1 & 2 years	Self-concept, locus of control, affect, stress, life satisfaction	Initial longitudinal analysis of change scores for the whole group showed that a shift from employment to unemployment or the reverse had little significant effect on psychological well-being. The shift from employment to unemployment led to more external causal attributions for youth unemployment; the reverse transition had the opposite effect. Sub-group analysis (O'Brien & Feather 1990) showed that those in good employment that let them utilize their skills and education had significantly higher personal competence, higher internal control, lower depressive affect and higher life satisfaction. Those in poor employment were more comparable to the unemployed. These results show that the relative effects of employment and unemployment depend on the quality of employment.

(Patton & Noller 1990)

TABLE 2. HEALTH IMPACTS OF EMPLOYMENT, RE-EMPLOYMENT AND RETIREMENT

Study	Population/setting	Follow-up	Health measures	Key findings on re-employment (*Additional reviewers' comments in italics*)

Table 2a: School leavers and young adults (Age < approx. 25 years) *continued*

Study	Population/setting	Follow-up	Health measures	Key findings on re-employment
(Patton & Noller 1990) Australia	216 school leavers	Before leaving school, 1 & 2 years	Self-image, Children's depression scale	There was little relation between baseline measures at school and subsequent employment (i.e. little mental health selection effect). Those who gained employment or returned to education showed slight but not significant improvement on all measures, while those who became unemployed deteriorated significantly on all measures. (*No separate data presented for those who moved from unemployment at time 2 to employment at time 3*).

TABLE 2. HEALTH IMPACTS OF EMPLOYMENT, RE-EMPLOYMENT AND RETIREMENT

Table 2a: School leavers and young adults (Age < approx. 25 years) *continued*

Study	Population/setting	Follow-up	Health measures	Key findings on re-employment (*Additional reviewers' comments in italics*)
(Tiggemann & Winefield 1984), (Winefield & Tiggemann 1990), (Winefield *et al.* 1990), (Winefield *et al.* 1991b) & (Winefield *et al.* 1991a) Australia	672 school leavers	Before leaving school, 1, 2, 3 & 7 years	Self-esteem, locus of control, -ve mood & depression scales	There was no relation between baseline measures at school and subsequent employment (i.e. no mental health selection effect). School leavers who entered employment or returned to full-time studies had increased internal locus of control, increased self-esteem, less negative mood and less depressive affect. However, those in unsatisfactory employment did much more poorly and were more comparable to the unemployed. Social support and financial security were the most important modifiers School leavers who became unemployed had lesser improvement in locus of control and self esteem and little change in mood or depressive affect. Those who subsequently became employed showed further gains to match those employed throughout. Longitudinal analysis suggested that the disadvantaged groups (unemployed or unsatisfactory employment) showed smaller improvements than the others (satisfactory employment or further studies), rather than actual deterioration. (*Combined results from various papers, based on subsets of same cohort*).

TABLE 2. HEALTH IMPACTS OF EMPLOYMENT, RE-EMPLOYMENT AND RETIREMENT

Table 2a: School leavers and young adults (Age < approx. 25 years) *continued*

Study	Population/setting	Follow-up	Health measures	Key findings on re-employment (*Additional reviewers' comments in italics*)
(Graetz 1993) (Morrell *et al.* 1994) Australia	2 samples: 9,000 & 2,403 16-25 year olds, excluding those with prior health problems	1, 2 & 3 years	GHQ, Psychiatric case rate	Students who entered the workforce had significant improvement in GHQ scores. Among school leavers, the beneficial effects of employment were much greater than the adverse effects of unemployment. Unemployed people who subsequently became employed had significant improvement in GHQ scores to normal levels and improvement in psychiatric case rate. Further analysis provided strong evidence that these health changes were a function of changes in employment status rather than predisposing health factors. The relative risk (RR) of becoming psychologically disturbed as a consequence of moving from employed to unemployed was 1.51 (95% CI 1.15-1.99) while the RR of those who were psychologically disturbed recovering upon re-employment was 1.63 (95% CI 1.08-2.48).
(Dooley & Prause 1995) US	2 samples: 3,369 & 5,969 school leavers	Before leaving school & 7 years	Self esteem, Abbreviated locus of control	School leavers who moved into satisfactory employment showed greatest average gains in self-esteem. Those who moved into unsatisfactory employment and those who remained unemployed showed lesser gains.

TABLE 2. HEALTH IMPACTS OF EMPLOYMENT, RE-EMPLOYMENT AND RETIREMENT

Table 2a: School leavers and young adults (Age < approx. 25 years) continued

Study	Population/setting	Follow-up	Health measures	Key findings on re-employment (*Additional reviewers' comments in italics*)
(Dooley & Prause 1995) US	2 samples: 3,369 & 5,969 school leavers	Before leaving school & 7 years	Self esteem, Abbreviated locus of control	School leavers who moved into satisfactory employment showed greatest average gains in self-esteem. Those who moved into unsatisfactory employment and those who remained unemployed showed lesser gains.
(Hammarström 1994a; Hammarström & Janlert 2002) Sweden	1,083 school leavers	Before leaving school, 5 & 14 years	Blood pressure, somatic and psychological symptoms, smoking, alcohol intake	Males and females who moved into and remained in stable employment had less increase in blood pressure and fewer somatic and psychological symptoms by age 21 (5 year follow-up) though all measures increased slightly by age 30 (14 year follow-up). Those who had early unemployment (>6 months unemployment between age 16-21) continued to have significantly increased somatic (men only) and psychological symptoms and smoked more (at 14 year follow-up), after controlling for initial health and social class. Alcohol consumption was unrelated to employment status or history.
(Mean Patterson 1997) UK	173 16-17 year-old unemployed	Baseline, 10-12 months	GHQ, Self-esteem	Employment led to lowering of GHQ: this was partly a selection effect but also causal. Employment did not produce any improvement in low self-esteem.

TABLE 2. HEALTH IMPACTS OF EMPLOYMENT, RE-EMPLOYMENT AND RETIREMENT

Table 2a: School leavers and young adults (Age < approx. 25 years) *continued*

Study	Population/setting	Follow-up	Health measures	Key findings on re-employment (*Additional reviewers' comments in italics*)
(Schaufeli 1997) Netherlands	a) 767 school leavers, b) 635 college graduates	Before final exam, a) 1 year b) 6, 12, 18 24 months	a) GHQ b) SCL-90 Various attitudes to work	School leavers who entered work or further studies showed slight improvement in GHQ, while those who became unemployed deteriorated markedly. Employment status had no significant effect on psychological distress in graduates (i.e. no causal effect). There was no mental health selection effect, but employment was predicted by a positive attitude and positive ways of coping with unemployment. These results may reflect the favourable Dutch structural and cultural context at the time.
Bjarnason & Sigurdardottir 2003) 5 Scandinavian countries + Scotland	7,300 18-24 year -old youths unemployed >3months	6 month follow up	Hopkins Symptom Checklist	Those who moved into permanent employment had much less psychological distress. Those who found temporary employment, returned to education or stayed at home also showed some lesser improvement. Perceptions of material deprivation and parental emotional support directly affected distress in all labour market conditions and mediated the influence of various other factors on distress.

TABLE 2. HEALTH IMPACTS OF EMPLOYMENT, RE-EMPLOYMENT AND RETIREMENT

Table 2b: Adults (Age ~25 to ~ 50 years)

Study	Population/setting	Follow-up	Health measures	Key findings on re-employment (*Additional reviewers' comments in italics*)
(Cohn 1978) US	National family panel sample	Baseline & 1 year	Self-satisfaction	Becoming unemployed leads to greater dissatisfaction with self. That decrement is reversed with re-employment, after taking into account residual effects of unemployment on familial role performance.
(Warr & Jackson 1985)	629 unemployed men	Baseline & 9 months	Reported general health, GHQ	On regression analysis, only duration of unemployment and age <60 years predicted re-employment. (*Various measures of baseline health and employment commitment were individually significant but no longer significant on multivariate analysis*). Those who regained paid work showed large improvements in all measures of health.
(Layton 1986b) UK	101 men facing compulsory redundancy. 62 re-employed	Baseline & 6 months	GHQ	Baseline GHQ did not predict re-employment (i.e. no selection effect). Those who were re-employed showed significant improvement in mental well-being, while those who remained unemployed showed a significant increase in minor psychiatric morbidity.
(Payne & Jones 1987)	399 unemployed men 203 middle class, 196 working class	Baseline, 12 months	GHQ, reported general health, financial worries	Re-employment led to improvement in general health, psychological distress, anxiety, depression and financial worries in both middle class and working class. Groups.

TABLE 2. HEALTH IMPACTS OF EMPLOYMENT, RE-EMPLOYMENT AND RETIREMENT

Table 2b: Adults (Age ~25 to ~50 years) *continued*

Study	Population/setting	Follow-up	Health measures	Key findings on re-employment (*Additional reviewers' comments in italics*)
(Vinokur *et al.* 1987) (Vinokur & Caplan 1987) US	297 unemployed 1 month & 189 employed males	Baseline, 4 & 12 months	Interview, Hopkins Symptom Checklist, self-esteem, life satisfaction	Childhood/adolescent stresses, Vietnam war stresses and unemployment had independent adverse effects on mental health. The adverse effects of unemployment were reversed by re-employment, in contrast to the adverse effects of childhood/adolescent and war stresses which were long-lasting. However, unsuccessful job seeking can have negative effects on mental health, though these effects can be counteracted by social support. These effects are particularly strong among highly motivated job seekers.
(Iversen & Sabroe 1988) Denmark	1,153 workers in shipyard closing down + 441 control workers. 374 moved from unemployment to re-employment	1, 2 & 3 years	GHQ	GHQ scores improved markedly and significantly. When unemployed, this group had comparable scores to other unemployed, when re-employed they had comparable scores to other employed people. This improvement remained significant after allowing for occupation, health and social support.

TABLE 2. HEALTH IMPACTS OF EMPLOYMENT, RE-EMPLOYMENT AND RETIREMENT

Table 2b: Adults (Age ~25 to ~ 50 years) *continued*

Study	Population/setting	Follow-up	Health measures	Key findings on re-employment (*Additional reviewers' comments in italics*)
(Vinokur *et al.* 1987) (Vinokur & Caplan 1987) US	297 unemployed 1 month & 189 employed males	Baseline, 4 & 12 months	Interview, Hopkins Symptom Checklist, self-esteem, life satisfaction	Childhood/adolescent stresses, Vietnam war stresses and unemployment had independent adverse effects on mental health. The adverse effects of unemployment were reversed by re-employment, in contrast to the adverse effects of childhood/adolescent and war stresses which were long-lasting. However, unsuccessful job seeking can have negative effects on mental health, though these effects can be counteracted by social support. These effects are particularly strong among highly motivated job seekers.
(Kessler *et al.* 1989) US	Stratified population sample 492 adults	Baseline & 1 year	Several measures of distress, self-assessed physical ill health	At baseline, all measures of distress were elevated in unemployed compared with stable employed. Distress at baseline did NOT reduce the probability of re-employment (i.e. no negative selection effect). Re-employment reduced levels of distress to those of stable employment, and led to complete emotional recovery within 1 year. However, a minority who were re-employed in insecure jobs had some persisting depression.

TABLE 2. HEALTH IMPACTS OF EMPLOYMENT, RE-EMPLOYMENT AND RETIREMENT

Table 2b: Adults (Age ~25 to ~ 50 years) *continued*

Study	Population/setting	Follow-up	Health measures	Key findings on re-employment (*Additional reviewers' comments in italics*)
(Caplan *et al.* 1989) (Vinokur *et al.* 1991a) (van Ryn & Vinokur 1992) US	928 recently unemployed adults	1 & 4 months	Quality of life, Hopkins Symptom Checklist	(*Randomized field experiment of training in job seeking, problem solving and reinforcement*). People who found re-employment scored significantly lower on anxiety, depression and anger and higher on self-esteem and quality of life. Effect sizes were 2–3X greater for those who actually participated on employment outcomes and mental health (anxiety and depression). Further analysis suggested that those who needed the intervention most and benefited from it were more likely to participate. Analysis based on the theory of planned behaviour demonstrated the mediational role of job-search self-efficacy and the long-term influence of inoculation against setbacks.
(Lahelma 1992) Finland	703 previously employed registered job seekers 25 to 49	Baseline , 3 months & 15 months	GHQ	Unemployment had an adverse impact on mental well-being; re-employment improved the mental well being of the unemployed. The results were similar at both follow-up points.

TABLE 2. HEALTH IMPACTS OF EMPLOYMENT, RE-EMPLOYMENT AND RETIREMENT

Table 2b: Adults (Age ~25 to ~ 50 years) *continued*

Study	Population/setting	Follow-up	Health measures	Key findings on re-employment (*Additional reviewers' comments in italics*)
(Virtanen 1993) Finland (Westin 1993) (Editorial)	84 teenage & 143 adult long-term unemployed, & 82 continuously employed	2 years	Frequency of primary health care visits	Re-employment was associated with an increased rate of consultation, from low to normal (compared with the continuously employed). Further unemployment was associated with a fall in consultation rates. (*These data are contrary to the usual findings that unemployment is associated with increased health care consumption. They may be associated with consultation for sick certification during a policy of enforced re-employment*).
(Hamilton *et al.* 1993) US	1597 autoworkers; 831 from closing & 766 from non-closing plants	3 months before closure, 1& 2 years	Hopkins Symptom Checklist	Unemployment at waves 2 and 3 was related to prior frequency of reported depressive symptoms. Distress improved in workers whose Wave 3 outcomes fitted their Wave 2 coping decisions: those who both wanted and found a job, lost a job they disliked, or remained unemployed as planned.

TABLE 2. HEALTH IMPACTS OF EMPLOYMENT, RE-EMPLOYMENT AND RETIREMENT

Table 2b: Adults (Age ~25 to ~50 years) *continued*

Study	Population/setting	Follow-up	Health measures	Key findings on re-employment (*Additional reviewers' comments in italics*)
(Claussen *et al.* 1993) (Claussen 1999) Norway	310 sample of 17-63 year-old registered unemployed >12 weeks 41% re-employed at 2-year follow-up	Baseline, 2- and 5-year	Psychometric testing, Hopkins symptom check list, GHQ, medical examination	There was considerable health related selection to re-employment, in terms of psychological distress and medical diagnosis: a psychiatric diagnosis was associated with a 70% reduction and normal psychometric tests with 2-3 times better chance of re-employment. The prevalence of mental disorders was reduced by getting a job, with significant improvements (compared with those who remained unemployed) in depression scores and psychiatric diagnosis. However, the improvement was less than in some other studies. There was no change in the somatic diagnosis rate (though those who remained unemployed showed an increase). At 5-years, re-employment reduced the prevalence of alcohol abuse by two-thirds (though there may have been some selection effect).

TABLE 2. HEALTH IMPACTS OF EMPLOYMENT, RE-EMPLOYMENT AND RETIREMENT

Table 2b: Adults (Age ~25 to ~ 50 years) *continued*

Study	Population/setting	Follow-up	Health measures	Key findings on re-employment (*Additional reviewers' comments in italics*)
(Burchell 1994) UK	365 adult unemployed: 200 men, 165 women	Baseline & 8 months	GHQ	Men: significant improvement in GHQ in those with secure re-employment; no change in those who with insecure re-employment or who remained unemployed. Women: significant but less marked improvement in GHQ in those with secure or insecure re-employment; no change in those who remained unemployed.
(Wanberg 1995) US	129 unemployed <4 months	Baseline & 9 months	Global and facet job satisfaction; life satisfaction; GHQ	The only significant main effects for time were found for mental health. Individuals moving from unemployment to satisfactory employment showed improved mental health. Those remaining unemployed or who moved into unsatisfactory employment showed no improvement in mental health. There was no significant change in global life satisfaction.
(Hamilton *et al.* 1997) Canada	Population sample E Montreal residents 330 employed and 350 unemployed	Baseline & 4 surveys over 14 months	Psychiatric Symptom Index	(*Complex econometric modelling & statistical analysis*) After controlling for endogeneity, better mental health improved employability and employment led to improved mental health.

TABLE 2. HEALTH IMPACTS OF EMPLOYMENT, RE-EMPLOYMENT AND RETIREMENT

Table 2b: Adults (Age ~25 to ~50 years) *continued*

Study	Population/setting	Follow-up	Health measures	Key findings on re-employment (*Additional reviewers' comments in italics*)
(Halvorsen 1998) Norway	1,000 sample of 20-59 year old long-term (>6 months) unemployed men and women	2 & 18 months	Various measures of psychological distress	(*Unusual measures of psychological distress + complex regression analyses*). The most important factors influencing psychological distress were financial hardship and marital breakup; increasing age and female gender also had some effect. Re-employment had little impact on psychological distress after confounding factors were allowed for. The most important characteristic of (re-) employment was whether it was secure. To some extent, distress was not due to unemployment per se, but due to a selection process ending in long-term unemployment experienced by people who were already distressed at 6 months. (These subjects were already long-term unemployed, therefore their mental health may have 'stabilised' before the study commenced).
(Nordenmark & Strandh 1999) Nordenmark 1999) Sweden	3,500 randomly selected unemployed	21 months	GHQ	Psychological distress improved with re-employment. It improved most in those with high economic and psychosocial needs. The results support the hypothesis that employment status determines both employment commitment and mental well-being.

TABLE 2. HEALTH IMPACTS OF EMPLOYMENT, RE-EMPLOYMENT AND RETIREMENT

Table 2b: Adults (Age ~25 to ~50 years) continued

Study	Population/setting	Follow-up	Health measures	Key findings on re-employment (Additional reviewers' comments in italics)
(Liira & Leino-Arjas 1999) Finland	781 construction & 877 forestry workers. All male and employed at baseline	5 years	Subjective health, musculoskeletal symptoms & distress	Previous history of unemployment predicted further unemployment. Re-employment led to improved distress and depressive symptoms (compared to those who became and remained unemployed), almost to level of whose who were continuously employed.
(Vuori & Vesalainen 1999) Finland	553 unemployed aged 18-54 years who had labour market interventions. 23% employed at follow-up	1 year	GHQ	Those who were re-employed showed significantly fewer symptoms, while those who remained unemployed showed no change in distress levels. There was no selection to re-employment according to distress levels. (Main focus was on predictors of outcome. No more detailed data presented on health impact of re-employment.)
(Ferrie et al. 2001) UK	666 civil servants in privatized department (513 men, 153 women)	18 months	GHQ, self-rated health, GP visits	Those who returned to secure employment had significantly better general health, less minor psychiatric morbidity and fewer GP consultations than those who remained unemployed. Those who permanently exited the labour force were comparable to those in secure employment. Those who returned to insecure employment were comparable to those who remained unemployed.

TABLE 2. HEALTH IMPACTS OF EMPLOYMENT, RE-EMPLOYMENT AND RETIREMENT

Table 2b: Adults (Age ~25 to ~ 50 years) *continued*

Study	Population/setting	Follow-up	Health measures	Key findings on re-employment (*Additional reviewers' comments in italics*)
(Ferrie *et al.* 2002) UK	3,685 civil servants who were still working after 10 years	2, 4, 7 & 10 years	General health, GHQ depression scores	Men and women who moved from insecure to secure employment had improvement in self-rated health, GHQ scores and depression scores (interpreted as 'minor psychiatric morbidity') compared to those who remained in insecure employment, although the GHQ and depression scores did not fully return to the levels of those who were in secure employment throughout.
(Ostry *et al.* 2002) Canada	3000 sawmill workers	1 year before de-industrialisation & 20 years	Self-reported health status	Long-term working conditions and health status were generally better for workers who, under pressure of de-industrialisation, left the sawmill industry and obtained re-employment outside this sector.
(Brenner 2002) 15 EU countries (including UK) + US	Aggregate time-series analysis from 1950-1998.	---	Mortality	Increased employment rates in the direction of a "full employment" economy are a fundamental source of decreased mortality. (*This analysis was the corollary of Brenner's early analyses of unemployment – see Table 1.*) National wealth (GDP per capita adjusted for inflation) is a simultaneous source of improvement in the health and longevity of populations.
(Thomas *et al.* 2005) UK	13,359 adults aged 16-74 years from British Household Panel Survey	8 waves 1991-98/99	GHQ	Transitions from employment to either unemployment or long-term sick leave were associated with significantly increased psychological distress for both men and women. Transitions into employment resulted in improvement in mental health, though the changes were generally not significant.

TABLE 2. HEALTH IMPACTS OF EMPLOYMENT, RE-EMPLOYMENT AND RETIREMENT

Table 2c: Older workers (> approx. 50 years) continued

Study	Population/setting	Follow-up	Health measures	Key findings on re-employment (Additional reviewers' comments in italics)
(Haynes et al. 1978) USA	2129 taking normal retirement at 65; 1842 taking early retirement (62-64)	5 years	Mortality	Mortality after early retirement is higher than would be expected in a corresponding working group: the only significant predictor is pre-retirement health status (i.e. a health selection effect). Taking all the data, no excess mortality was observed after normal retirement.
(Ekerdt et al. 1983) USA	263 male retirees (from long-term community survey) with pre- and post-retirement data	3 years	Perceived health effects of retirement	Compared 114 men who claimed retirement had a good effect on health with 149 men who claimed no effect. Retrospective claims of good effects were not corroborated by longitudinal pre- to post-retirement improvement in self-reported health. Claims of good effect were more likely among men who, prior to retirement, had serious health problems or had jobs that required performing under stress. The extent to which health really improves upon retirement is a matter of perspective (subjective vs. objective measures of health).

TABLE 2. HEALTH IMPACTS OF EMPLOYMENT, RE-EMPLOYMENT AND RETIREMENT

Study	Population/setting	Follow-up	Health measures	Key findings on re-employment (*Additional reviewers' comments in italics*)
Table 2c: Older workers (> approx. 50 years) *continued*				
(Frese & Mohr 1987) (Frese 1987) Germany	51 unemployed blue-collar workers aged >45 years	Baseline & 18 months	Financial problems, general activity level, depression, control	Those who were re-employed or who took early retirement showed improvements in financial problems and depression, while those who remained unemployed or who were temporarily re-employed and then unemployed again showed deterioration in financial problems, loss of control and depression. People who retired out of unemployment improved in depression similar to those who found a job. Disappointed hope (to find a job) can lead to depression.
(Reitzes *et al.* 1996) USA	438 employed and 299 retired persons age 58 to 64 at baseline: 52% female	2 years	Self-esteem, depression	An investigation of changes in self-esteem and depression as older workers either move into retirement or continue their full time employment. Self-esteem scores did not change for either group during follow-up, but depression scores declined for workers who retired. Comparing retirees with those who continued to work, there was no evidence of a negative retirement effect. It does not appear that workers enter retirement with well-being scores any different to workers who continue work. Worker identity meanings become less relevant to individuals as they move into retirement, but for those who continue to work a positive set of previous worker-identity meanings helps to resist or lower depression.

TABLE 2. HEALTH IMPACTS OF EMPLOYMENT, RE-EMPLOYMENT AND RETIREMENT

Table 2c: Older workers (> approx. 50 years) *continued*

Study	Population/setting	Follow-up	Health measures	Key findings on re-employment (*Additional reviewers' comments in italics*)
(Salokangas & Joukamaa 1991) Finland	389 people age 62 at baseline. 80 were on disability pension; 75 on occupational pension; 184 still working	4 years	Physical health (perceived) + mental health (GHQ)	Retirement itself had no great immediate effects on physical health in general. Mental health of subjects who retired at normal old-age retirement (65 y) became better than that of subjects who retired before the study (i.e. those on disability or occupational pension) – this is in accordance with the view that retirement is a positively anticipated event for working people, and may be related to the relief of work stress and sense of fulfilment by having worked up to normal age.
(Morris *et al.* 1992) (Morris *et al.* 1994) UK	Men aged 40-59 who had been continuously employed for 5 years; 1,779 experienced unemployment or retired; 4,412 continuously employed	Baseline & 5 years	Weight gain, smoking, alcohol intake. Mortality	Men who remained continuously employed had the lowest mortality, even after adjusting for socio-economic variables, manual/non-manual work and health-related behaviour. Even men who retired for reasons other than illness and who appeared to be relatively advantaged and healthy had a significantly increased risk (RR 1.87). The effect was non-specific: the increased risk of mortality from cancer was similar to that from cardiovascular disease. Men who remained in continuous employment maintained their body weight; men who experienced any form of non-employment were significantly more likely to lose or gain >10% in body weight. There was no evidence that non-employment led to increased smoking or alcohol intake. Men non-employed because of illness were significantly more likely to reduce their smoking and alcohol intake.

TABLE 2. HEALTH IMPACTS OF EMPLOYMENT, RE-EMPLOYMENT AND RETIREMENT

Table 2c: Older workers (> approx. 50 years) continued

Study	Population/setting	Follow-up	Health measures	Key findings on re-employment (Additional reviewers' comments in italics)
(Gallo et al. 2000) (Gallo et al. 2001) US	3,119 older workers aged 51-61 years - 209 had involuntary job loss	Baseline & 2 years	Physical functioning (activities of daily living), CES-Depression scale	Involuntary job loss had a significant negative effect on physical functioning and mental health, even after allowing for baseline health status and socio-demographic factors. Results demonstrate both selection and causal mechanisms. Older and unmarried people may be especially vulnerable. Re-employment was significantly, positively associated with physical functioning and mental health at the follow-up interview. (Results presented as regression analyses and no descriptive statistics, so not possible to assess how fully the effects were reversed).
(Quaade et al. 2002) Denmark	Birth cohorts 1926-36 (age 50-60 at baseline)	1987 - 1996	Mortality	In Denmark disability benefit was granted mainly for poor health; early retirement benefit was earned through long-term membership of an unemployment benefit scheme. Standardised mortality rates were low in employed persons (0.59 and 0.51 for men and women), high in disability beneficiaries (2.31 and 1.66), and in-between for early retirement benefit recipients (0.88 and 0.72). Disability benefit recipients had a high mortality immediately after retirement, probably due largely to their pre-existing health condition. In early retirement recipients the relative risk of death increased with time since retirement, consistent with an adverse effect on health of retirement itself. However, it was difficult to disentangle the effects of retirement from the effects of earlier health-associated selection into the two schemes.

TABLE 2. HEALTH IMPACTS OF EMPLOYMENT, RE-EMPLOYMENT AND RETIREMENT

Table 2c: Older workers (> approx. 50 years) *continued*

Study	Population/setting	Follow-up	Health measures	Key findings on re-employment (*Additional reviewers' comments in italics*)
(Mein *et al.* 2003) UK	392 retired and 618 working civil servants aged 54 to 59 at baseline	Average 36 months	SF-36 health functioning; employment grade	Mental health functioning deteriorated over time among those who continued to work, and improved among those who retired. However, improvements in mental health were restricted to those in higher employment grades. Physical health functioning declined over time in both working and retired civil servants. In summary, retirement at age 60 had no effect on physical health functioning and, if anything, was associated with an improvement in mental health, particularly among high socioeconomic status groups. (*Data from British civil servants may not be generalisable to other occupational groups*).
(Gallo *et al.* 2004) US	4,220 older workers aged 51–61 years - 457 had involuntary job loss	Baseline & 2, 4, 6 years	Myocardial infarction Stroke	Those who remained in continuous employment had less than half the relative risk of stroke compared with those who had late-career involuntary job loss. There was statistically significant effect on myocardial infarction. (*No re-employment data were presented*).

TABLE 2. HEALTH IMPACTS OF EMPLOYMENT, RE-EMPLOYMENT AND RETIREMENT

Table 2c: Older workers (> approx. 50 years) *continued*

Study	Population/setting	Follow-up	Health measures	Key findings on re-employment (*Additional reviewers' comments in italics*)
(Tsai *et al.* 2005) US	Oil company employees who retired at 55, 60 or 65; employees who continued working (n=3,668)	20+ years	Mortality	After adjusting for socioeconomic status, employees who retired early at 55 had greater mortality than those who retired at 65 – the mortality was about twice as high in the first 10 years after retirement. Early retirees who survived to 65 had higher post-65 mortality than those who had continued working. Mortality was similar in those who retired at 60 and 65. Mortality did not differ for the first 5 years after retirement at 60 compared with continuing work.
(Thomas *et al.* 2005) UK	British Household Panel Survey	8 waves 1991-98/99	GHQ	Transitions from employment to retirement and vice versa were associated with non-significant changes in psychological distress in men or women. (*Note this study also appears in Adults section*).

TABLE 2. HEALTH IMPACTS OF EMPLOYMENT, RE-EMPLOYMENT AND RETIREMENT

Study	Population/setting	Follow-up	Health measures	Key findings on re-employment (*Additional reviewers' comments in italics*)
Table 2c: Older workers (> approx. 50 years) *continued*				
(Pattani *et al.* 2004) UK	1317 ill retirees in NHS pension scheme (27% < 50)	Baseline, 12 months	Quality of life, SF-36	Retirees' quality of life improved, but at 1-year remained lower than the general population. 13% were working again, mostly part-time. Those who were working showed greater improvements in physical and mental component scores, compared with those who were not working.

TABLE 3: WORK FOR SICK AND DISABLED PEOPLE

Authors	Key features (*Additional reviewers' comments in italics*)

Table 3a: Disability

Authors	Key features
(United Nations 1948)	**Universal Declaration of Human Rights** Article 23.1 Everyone has the right to work, to free choice of employment, to just and favourable conditions of work and to protection against unemployment.
(United Nations 1975)	**Declaration on the rights of disabled people** Article 4: Disabled persons have the same civil and political rights as other human beings. Article 7: Disabled persons have the right to economic and social security and to a decent level of living. They have the right, according to their capabilities, to secure and retain employment or to engage in a useful, productive, and remunerative occupation.
(Council of Europe 1992) Recommendation No. R(92)6	**Policy for people with disabilities** Principle 2: policy should aim at guaranteeing full and active participation in community life. All people who are disabled **or are in danger of becoming so** (*our emphasis*) – should – – have as much economic independence as possible, particularly by having an occupation as highly qualified as possible and a commensurate personal income.
(ILO 2002) Code of Practice	**Managing disability in the workplace** (International Labor Organisation) This Code is based on the principles underpinning international instruments and initiatives designed to promote the safe and healthy employment of all persons with disabilities. (*The Code is implicitly based on principles of full participation, equality of opportunity, non-discrimination and accommodation as being in the best interests of disabled people. However, there is no explicit statement of these principles nor any evidence on the health benefits for disabled people.*) The objectives of the code are to provide practical guidance on the management of disability issues in the workplace with a view to: (a) ensuring that people with disabilities have equal opportunities in the workplace; (b) improving employment prospects for persons with disabilities by facilitating recruitment, return to work, job retention and opportunities for advancement; (c) promoting a safe, accessible and healthy workplace; (d) assuring that employer costs associated with disability among employees are minimized – including health care and insurance payments, in some instances; (e) maximizing the contribution which workers with disabilities can make to the enterprise.

TABLE 3: WORK FOR SICK AND DISABLED PEOPLE

Authors	Key features (*Additional reviewers' comments in italics*)

Table 3a: Disability *continued*

(IoD 2002) Guidance	**Health and wellbeing in the workplace** (Institute of Directors, UK) Lays out the business case for promoting and maintaining health, safety and well-being at work. The most important asset of any organisation, whatever its size, is its people: healthier staff contribute more to corporate success. Offers directors and employers practical advice on how to identify which health and workplace issues are specific to their workplace and outlines measures to help address them. Covers current workplace health issues (stress and musculoskeletal disorders); health risk assessment, management and insurance; flexible working; employee support; sickness absence management; vocational rehabilitation; developing health policy; and legislative changes.
(The Council of the European Union 2003) Resolution of 15 July 2003	**Policy for people with disabilities** Principle VII.1.1 Employment: To permit the fullest possible vocational integration of people with disabilities, whatever the origin, nature, and degree of their disability, and thereby also to promote their social integration and personal fulfilment, all individual and collective measures should be taken to enable them to work, whenever possible in an ordinary working environment, either as a salaried employee or self-employed person. The Council of the European Union – – calls on the Member States and the Commission, – – to continue efforts to remove barriers to the integration and participation of people with disabilities in the labour market– –.
(OECD 2003) Report	Transforming Disability into Ability: Policies to Promote Work and Income Security for Disabled People (Organisation for Economic Co-operation and Development) Personal incomes of disabled people depend primarily on their work status. Average work incomes of those disabled people who have a job are almost as high as average work incomes of people without disabilities. Disabled people without a job have considerably lower personal financial resources. One of the two goals of disability policy is to ensure that disabled citizens are not excluded: that they are encouraged and empowered to participate as fully as possible in economic and social life, and in particular to engage in gainful employment, and that they are not ousted from the labour market too easily or too early. (*The other goal of disability policy is to provide income security*).

TABLE 3: WORK FOR SICK AND DISABLED PEOPLE

Authors

Key features *(Additional reviewers' comments in italics)*

Table 3a: Disability *continued*

Authors	Key features
(Disability Rights Commission 2004) Policy paper	**Disability Rights Commission Strategic Plan 2004–2007** The goal of the UK Disability Rights Commission is: "a society where all disabled people can participate fully as equal citizens". "All our work will be geared to 'closing the gap' of inequality between disabled and non-disabled people. That means the gap in educational attainment, in career opportunities (not just jobs), earned income – – so that disabled people of all ages can live independently and contribute fully to their communities."
Various-summarised in: (Waddell & Aylward 2005)	**Summary of UK Disability Literature** 50% of 'disabled' people do work and 34% of people on disability and incapacity benefits say they want to work. The medical evidence is that many people with longer-term sickness still have (some) capacity for (some) work despite their health condition. Incapacity benefits have many financial and other disadvantages compared with work. On average, disabled people in work earn 80-90% as much as non-disabled workers, while disabled people who do not work earn less than half that amount (OECD 2003; DRC 2004b). For these reasons, work is the best exit from incapacity benefits and, in that sense, 'work is the best form of welfare' (Mead 1997; Field 1998; King & Wickam-Jones 1999). The major provisos are: • For those who can; this must not become an excuse for forcing people off benefits when they cannot reasonably be expected to work. • That suitable work and opportunities are available in the (local) labour market. • That adequate and effective support into work is provided. Most people – including disabled people (Oliver & Barnes 1998; DRC 2004a; Howard 2004) and Incapacity Benefit recipients (Goldstone & Douglas 2003) – agree that whenever possible productive employment is preferable for sick and disabled people, their families and society at large, rather than relying on financial benefits as an incomplete replacement for income. There is strong public support for encouraging benefit claimants back to work, when this is feasible (Williams *et al.* 1999; Saunders 2002a). However, the preference is for an approach that encourages and helps people to work rather than compels them to do so. Many benefit recipients agree but point out that the onus is on society to make decent, adequately paid jobs available, and that the process and implementation of any 'encouragement' must be fair and reasonable and allow for individual circumstances (Dwyer 2000; DRC 2004c).

TABLE 3: WORK FOR SICK AND DISABLED PEOPLE

Authors	Key features (*Additional reviewers' comments in italics*)

Table 3b: Sickness absence and return to work *continued*

(Schwefel 1986) Narrative review	**Unemployment, health and health services in German-speaking countries**
	There are manifold links between (un)employment and (ill) health: Employment may lead to illness because of unfavourable working conditions or job insecurity. Illness may lead to sickness absence and increased risk of losing employment. Unemployment may aggravate existing, or lead to new, physical or mental ill-health, and reduce the chances of re-employment (the average duration of unemployment in people with health problems is 3x that of healthy unemployed). Alternatively, however, stopping work may improve health. These links may be modified by the personal characteristics of the (un)employed and by various social factors. The main determinant of re-entry to work is the state of the labour market; the other major barriers include age, gender, qualifications and job history; health problems are usually not the major barrier. Unemployment often ends by moving into other social roles: long-term sickness, retired, or housewife.
(Kazimirski 1997) Policy Statement	**The physician's role in helping patients return to work after an illness or injury.** (Canadian Medical Association)
	Starts from the premise that 'Prolonged absence from one's normal roles, including absence from the workplace, is detrimental to a person's mental, physical and social well-being. Physicians should therefore encourage a patient's return to function and work as soon as possible after an illness or injury, provided that return to work does not endanger the patient, his or her co-workers or society. A safe and timely return to work benefits the patient and his or her family by enhancing recovery and reducing disability.' (*However, no evidence is presented or referenced to support this*).
	The physician's role is to diagnose and treat the illness or injury, to advise and support the patient, to provide and communicate appropriate information to the patient and employer and to work closely with other health professionals to facilitate the patient's safe and timely return to the most productive employment possible.
(BSRM 2000) Report	**Vocational Rehabilitation** (British Society of Rehabilitation Medicine)
	Definition: 'vocational rehabilitation is a process whereby those disadvantaged by illness or disability can be enabled to access, maintain or return to employment, or other useful occupation'.
	The report focuses on the early management of disability due to illness or injury. It concludes that better, speedier, more focused management of sickness absence with the aim of job retention and earlier return to work will - - -improve the quality of life for those involved. Successfully rehabilitated individuals feel confident about their work abilities and general well-being.

TABLE 3: WORK FOR SICK AND DISABLED PEOPLE

Authors	Key features (*Additional reviewers' comments in italics*)

Table 3b: Sickness absence and return to work *continued*

Authors	Key features
(AAOS 2000) Position Statement	**Early return to work programmes** (American Academy of Orthopedic Surgeons) 'As patient advocates, physicians realise that early return to work has many benefits for the injured worker.' 'The AAOS believes that safe early return to work programmes are in the best interests of patients. Studies have demonstrated that prolonged time away from work makes recovery and return to work progressively less likely. Return to work in light duty, part time or modified duty programmes is important in preventing the deconditioning and psychological behaviour patterns that inhibit successful return to work and in improving quality of life for the injured worker.'
(CBI 2000) Report	**Occupational health partnerships** (Confederation of British Industry) Stresses the importance of a healthy workforce and work place, the need to improve UK occupational health and rehabilitation services, that businesses of all sizes understand the need to manage areas such as stress and rehabilitation, and the need for Government, employers, unions and workers to work together.
(TUC 2000), (TUC 2002) Report	**Rehabilitation – consultation document and report** (Trades Union Congress, UK) 'People who are injured or made ill by their work should have the right to return to health and return to work so that they can maintain their earnings and continue their careers. The TUC is committed to improving existing provision and stimulating new ways to get people better and get people back to work.' 'The challenge of rehabilitation is a moral and humanitarian one, but there is also an economic imperative at work here.'
(BICMA 2000) Code of best practice	**Rehabilitation, early intervention and medical treatment in personal injury claims** (Bodily Injury Claims Management Association, UK) 'Employment issues – – can be addressed for the benefit of the claimant to enable him or her to keep the job that they have, to obtain alternative suitable employment with the same employer, or to re-train for new employment. – – if these needs are addressed at the proper time the claimant's quality of life and long term prospects may be greatly improved.'

TABLE 3: WORK FOR SICK AND DISABLED PEOPLE

Authors	Key features (*Additional reviewers' comments in italics*)

Table 3b: Sickness absence and return to work *continued*

Authors	Key features
(ABI/TUC 2002) Discussion paper	**Getting back to work** (Association of British Insurers / Trades Union Congress, UK) For the injured worker and their family: • Medical recovery can be accelerated and enhanced by an assisted return to the workplace programme • Successful rehabilitation would improve their long-term prospects in terms of physical and mental well-being, quality of life, employment and reintegration into society.
(IUA 2003) Revised Code & Guide	**Rehabilitation Code of Best Practice and Guide to Rehabilitation** (International Underwriters Association/Association of British Insurers/Bodily Injury Claims Management Association, UK) The Rehabilitation Code is based on the principle that most injured people want, first and foremost, to make an optimum and speedy recovery; and that their medical, psychological, social, and practical needs should be considered as soon as possible. The aim is to promote the use of rehabilitation and early intervention in the claims process so that the injured person makes the best and quickest possible medical, social, and psychological recovery.
(ACOEM 2002) Consensus Opinion Statement	**The attending physician's role in helping patients return to work after an illness or injury** (American College of Occupational and Environmental Medicine) '-- prolonged absence from one's normal roles, including absence from the workplace, is detrimental to a person's mental, physical and social well-being -- . A safe and timely return to work benefits the patient and his or her family by enhancing recovery, reducing disability, and minimizing social and economic disruption. -- The attending physician's role is (therefore) to -- facilitate the patient's safe and timely return to the most productive employment possible'. (*Adapted from Kazimirski 1997*).

TABLE 3: WORK FOR SICK AND DISABLED PEOPLE

Authors	Key features (*Additional reviewers' comments in italics*)

Table 3b: Sickness absence and return to work *continued*

Authors	Key features
(James *et al.* 2002), (James *et al.* 2003) HSE Research Report	**Job retention and vocational rehabilitation: development and evaluation of a conceptual framework** In the UK, the Health and Safety Commission and Health and Safety Executive have been paying increasing attention to the question of what can be done to increase the likelihood that employees who are sick or injured are able to be retained in employment and returned to their jobs, or, failing this, are able to obtain alternative employment with the same or another employer. A proactive approach to facilitating the (early) return to work and the continued employment of ill and injured workers can have benefits for individuals and organisations. This project developed a short 'framework' document about the processes and practices that are central to vocational rehabilitation and successful job retention. There is widespread agreement among all stakeholders that early and timely intervention exerts a crucial influence over rehabilitation outcomes: it can help to minimise the emotional detachment and associated mental health conditions that can develop among absent workers and prevent acute conditions becoming chronic.
(Kendall 2003) Unpublished Report for Department for Work and Pensions, UK	**Evidence Review: Raising the awareness of key frontline health professionals about the importance of work, job retention, and rehabilitation for their patients** The review considered what the current evidence on the best way to raise awareness of key front-line health professionals about the importance of work, job retention, and rehabilitation to their patients. It concluded that (*conclusions rearranged*): • There is an important and complex relationship between employment status and health, supported by a strong evidence base. There is a causal link between unemployment and deterioration in health status, and there is also a selection process so that people with health problems have more trouble getting paid jobs and sustaining employment. The effect of employment status on health is greater than the effect of health on employment status. The effects of several moderating variables including age, gender, migrant status, and duration complicate the relationship between employment status and health. • Participation in productive activity has a wide range of physical and psychological benefits to individuals, their families, and society. Work is at the very core of contemporary life for most people, providing financial security, personal identity, and an opportunity to make a meaningful contribution to community life. (*This appears to be a philosophical and theoretical argument rather than evidence-based in Kendall's review.*)

TABLE 3: WORK FOR SICK AND DISABLED PEOPLE

Authors	Key features (*Additional reviewers' comments in italics*)

Table 3b: Sickness absence and return to work *continued*

(Kendall 2003) Unpublished Report for Department for Work and Pensions, UK	**Evidence Review: Raising the awareness of key frontline health professionals about the importance of work, job retention, and rehabilitation for their patients** (*continued*)
	
	• Productive activity has a role in determining both short-term and long-term health. The negative consequences of economic inactivity, unemployment, and/or underemployment include:
	
	  • Physical health - elevated risk of specific diseases, suppressed immunological function (not clear how much evidence), or early death
	
	  • Psychological health – elevated risk of general distress or specific disorders including depression, anxiety, somatisation, or suicidal behaviour
	
	  • Health behaviours — elevated risk of commencing tobacco smoking (mixed evidence), increased use of health care services
	
	• Some individuals who are prescribed sickness absence are at risk of losing work habits, motivation, and work relationships. This places them at risk of not returning to work and therefore exposes them to important health risks.
	
	• Many health professionals have insufficient training and lack awareness or interest in occupational issues and outcomes.
	
	• There is an implicit tension in the role of health care professionals determining work capability and facilitating the return to work process.
	
	• Clinicians sometimes behave defensively and put people off work for extended periods (increasing the risk of health effects due to economic inactivity) because they fear being blamed for any symptom exacerbation.
	
	• There is considerable variation in rates of initiating time off work due to illness or injury, duration of sick leave, and the provision of effective return to work pathways.
	
	• The principles of best clinical practice should include:
	
	  • Avoiding the assumption that the workplace is a harmful environment when a person has illness or injury
	
	  • An ill or injured person should not automatically be advised time off work (unless there is a significant safety factor, or public health issue such as infection) (*continued*)

TABLE 3: WORK FOR SICK AND DISABLED PEOPLE

Authors	Key features (*Additional reviewers' comments in italics*)
Table 3b: Sickness absence and return to work *continued*	

(Kendall 2003) Unpublished Report for Department for Work and Pensions, UK	**Evidence Review: Raising the awareness of key frontline health professionals about the importance of work, job retention, and rehabilitation for their patients** (*continued*)
	
	• Before moving to full time off work status the option of undertaking selected duties should be appropriately and fully explored
	  • The best way to avoid the need to return to work is to remain at work.
	  • If the decision is made to put a person off work, this should be accompanied at the same time by a clear plan on how and when to return that person to work.
	• Clinicians should be encouraged to assume a safe and sustainable return to productive activity will convey the best health benefit to their patient, rather than assuming that absence from work will enhance outcomes.
	• Retention of existing employment is preferable to either unemployment, or seeking work with a different employer. Using the existing workplace as an integral part of the rehabilitation process is more effective than viewing it as a place to return an injured or sick employee to following completion of (medical) rehabilitation
	• It is critical that unemployment does not become medicalised, since this may do more harm than good to all involved, and breaches the fundamental health professional principle of 'first, do no harm'. People entering the healthcare system for whatever reason should not be exposed to the negative health consequences of extended incapacity or unemployment as a consequence of clinical practices.

(NIDMAR 2004) Code of practice	**Disability management** (National Institute of Disability Management and Research, Canada) Provides a framework within which employers, unions, legislators, insurers and providers can work together to support return to work for workers with disabilities. Identifies best management practices and policies for sound workplace programmes in disability management. The main 'values' are: • Safe and productive employment of workers with disabilities. • Safe and healthy working • Reduced occurrence and impact of illness and injury due to work • Consensus among government, labour and management on the achievement of these values Disability management requires the coordination of health care and support services, protection of confidentiality and informed consent, return to work planning, coordination of financial resources and information, occupational health and safety, dispute resolution procedures, education of all parties. Central to the approach is to remove obstacles within the workplace, policy and regulations. Lays out the responsibilities of the key participants in disability management, including a return to work coordinator / disability management professional.

TABLE 3: WORK FOR SICK AND DISABLED PEOPLE

Authors	Key features (*Additional reviewers' comments in italics*)
Table 3b: Sickness absence and return to work *continued*	
(Hemp 2004) Narrative review	**Presenteeism: at work but out of it** 'Presenteeism' refers to employees who are at work but not as productive as usual because of health problems. Presenteeism is usually assessed by questionnaires asking workers about times when they have attended work even though they had (an episode of) a health condition that they felt should require sick leave. It is difficult to quantify because of reliance on self-report of (a) ill health and (b) the impact it has on individual productivity, but is generally considered to carry very substantial costs for employers. Presenteeism is most often attributed to common health problems such as headache, cold/flu, asthma, allergies, fatigue/depression, stress, digestive problems, and musculoskeletal disorders. Solutions are seen to lie in the areas of awareness and health education (for employees and employers), improved illness management, provision of appropriate health care (including pharmacological), and compliance with health care advice and treatment. *(Presenteeism appears to be primarily a business rather than a health care concern: it is considered mainly in the occupational and business literature rather than the clinical literature, and focuses on health-related productivity losses. There is little or no suggestion that work either caused the health problem or (generally) will make it worse. The emphasis is generally on recognizing and managing the problem in the workplace, rather than arguing for keeping workers off work).*
(Vingård et al. 2004b) Narrative review	**Sickness presence** The term 'sickness presence' (analogous to presenteeism) is used here to describe situations where the ability to work is impaired due to disease, but yet the person goes back to work. A problem is that the term 'sickness presence' implies that being present at work is something exceptional if a person is sick. Most people diagnosed with a disease or disorder do, however, go to work and are not sick listed. Furthermore, the term is rather diffuse, and it would be beneficial if one or more specific terms could be used. The current body of scientific literature does not provide sufficient evidence to draw conclusions on the consequences of sickness presence.

TABLE 3: WORK FOR SICK AND DISABLED PEOPLE

Authors	Key features (*Additional reviewers' comments in italics*)

Table 3b: Sickness absence and return to work *continued*

Authors	Key features
(Vingård *et al.* 2004a) Systematic review	**Consequences of being on sick leave** Theoretically, sick leave may have positive and/or negative consequences for the sick-listed individual, e.g. regarding disease, physical and mental health, working life, lifestyle and quality of life. However, the circumstances and duration of sick leave varies greatly, and it is difficult to separate the consequences of sick leave per se from the consequences of the illness which led to sick leave. The few studies on the health consequences of sickness absence are mainly of low quality and provide insufficient evidence to permit any firm conclusions. There is limited evidence that long-term sick leave is associated with lower subsequent earnings. The few studies on the health consequences of disability pension are mainly of low quality and provide insufficient evidence to permit any firm conclusions. There is limited evidence that significant minorities (of one quarter to one third) of disability pension recipients' report that their quality of life had improved or deteriorated. (*The most striking finding of this review was the lack of relevant studies*).

TABLE 3: WORK FOR SICK AND DISABLED PEOPLE

Authors   Key features *(Additional reviewers' comments in italics)*

Table 3b: Sickness absence and return to work *continued*

(DWP 2004b)
Guidance

Sick certification: a guide for registered medical practitioners.
(Department for Work and Pensions, UK)

'Advice regarding fitness for work is an everyday part of the management of clinical problems and doctors should always consider carefully whether advice to refrain from work represents the most appropriate clinical management. Doctors can often best help a patient of working age by taking action which will encourage and support work retention and rehabilitation. When providing advice to a patient about fitness for work you may wish to consider the following factors:

• The nature of the patient's medical condition and how long the condition is likely to last.

• The functional limitations which result from the patient's condition, particularly in relation to the type of task they actually perform at work

• Any reasonable adjustments that might enable the patient to continue working

• Any appropriate clinical guidelines (e.g. for low back pain)

• Clinical management of the condition which is in the patient's best interest regarding work fitness, including managing the patient's expectations in relation to their ability to continue working.

In summary, you should always bear in mind that a patient may not be well served in the longer term by medical advice to refrain from work, if more appropriate clinical management would allow them to stay in work or return to work.'

'Help to return to work: It is recognised that the opportunity to do some work can help to improve a patient's condition and hopefully lead eventually to a return to regular employment. - - In some cases where the patient's condition could lead to prolonged sickness absence, you may wish to seek early specialist help - - - There are a range of local support services, including those available through Jobcentre Plus, which may be available to a patient who is not working, or at risk of losing their job, because of a medical condition or disability.'

TABLE 3: WORK FOR SICK AND DISABLED PEOPLE

Authors	Key features (*Additional reviewers' comments in italics*)

Table 3b: Sickness absence and return to work *continued*

Authors	Key features
(DWP 2003) Guidance	**Patients, their employment and their health: how to help your patients stay in work** (Department for Work and Pensions, UK) GP advice is important in shaping patient and employer beliefs and influencing return to work • As time away from work goes on significant heal effects can occur – depression and other psychological problems increase whatever the original diagnosis. • The longer a patient is off work the lower the chances of returning. Less than 50% of people with 6 months sickness absence ever return to work and few people return to any form of work after 1-2 years absence, irrespective of further treatment. • Strategies directed towards job retention are of proven value: they are needed in the first months of sickness absence. • Return to work after acute symptoms of depression have eased, but before it has completely resolved, may aid recovery. Support work resumption: • Suggest work adjustments, where appropriate, rather than signing the patient off work. • Suggest work adjustments if the patient is off sick – to enable early return to work. • Prescribe graduated work and/or transitional arrangements. • Suggest workplace assessment by workplace occupational health professional. • Dispel the myth that employers cannot dismiss the patient while off sick. If other (*non-work or non-health*) factors are the main problem, getting back to work may aid recovery. Signing patients off work may risk their job and add to their problems. Say as you advise the patient: 'There will come a point at which work will make you feel better – we don't want to miss that'.

TABLE 3: WORK FOR SICK AND DISABLED PEOPLE

Authors	Key features *(Additional reviewers' comments in italics)*
Table 3b: Sickness absence and return to work *continued*	
(Waddell & Burton 2004) Theoretical & conceptual review	**Concepts of rehabilitation for the management of common health problems** Develops five fundamental concepts: • 'Common health problems' – mild/moderate mental health, musculoskeletal and cardio-respiratory conditions that now cause about two-thirds of longer-term sickness absence, incapacity for work and early retirement on health grounds. Epidemiology shows that similar symptoms are very common in the general population, but they do not necessarily mean 'illness' and most people cope with them or recover uneventfully. There is often little or no objective pathology or severe and permanent impairment, and long-term incapacity is not inevitable. There are usually associated psychosocial problems. The less the impairment and the more subjective the complaints, the more central the role of personal/psychological factors. • 'Obstacles to recovery' – health-related, personal/psychological and social/occupational factors that aggravate and perpetuate disability and, critically, may also act as obstacles or barriers to normal recovery and (return to) work after injury or illness. • Rehabilitation for common health problems is about identifying and addressing obstacles to recovery. Rehabilitation principles should be an integral part of good clinical and occupational management. • The evidence shows that the best 'window of opportunity' for effective rehabilitation is between about 1 and 6+ months off work (though the exact limits are unclear). • Every health professional that treats patients with common health problems should be interested in and take responsibility for rehabilitation and occupational outcomes. This does not mean that every health professional should become a 'rehabilitation specialist'; rather, it goes to the heart of what good clinical management is all about. *(Starts from the premise that 'there is now broad agreement on the importance of rehabilitation and the need for better occupational health and vocational rehabilitation services in the UK' and the assumption that this will improve health and vocational outcomes for sick and disabled people, but does not provide any direct evidence on the health benefits of (return to) work) (Also in Table 5).*

TABLE 3: WORK FOR SICK AND DISABLED PEOPLE

Table 3b: Sickness absence and return to work continued

Authors	Key features (Additional reviewers' comments in italics)
(DWP 2004a) Discussion paper (Coleman & Kennedy 2005) Research report	**Framework for vocational rehabilitation** (Department for Work and Pensions, UK) From responses to the Government's policy paper and qualitative research, it is clear that many UK stakeholders (including employers, insurers, disability groups, health professionals and academics are 'enthused' about the potential of vocational rehabilitation.
(HSE 2004b) Guidance	**Managing sickness absence and return to work: an employers' and managers' guide** (Health and Safety Executive, UK) The guidance starts from several key standpoints: • Work, provided it is managed safely and effectively, is essential to good health and well-being • Inability to get back to work due to poor health brings on more health problems, both physical and mental • Sickness absence is a major cost to industry – effective management makes good business sense • Early management of sickness absence is essential to prevent long-term absence The guide describes the best practice steps for managing sickness absence, and promotes the concept of 'recovery of health at work'. It covers a wide range of issues: • Importance and understanding of the issues • Legal obligations and responsibilities • Managing recovery at work • Recording sickness absence • Keeping in contact • Return to work interview • Planning workplace adjustments • Making use of professional and other advice and treatment • Agreeing and reviewing a return to work plan • Coordinating the return to work process • Developing and implementing a sickness absence and return to work policy. *(This was the first major guidance from the Health & Safety Executive on sickness absence management, based on a mix of evidence-based and consensus-based best practice. It notes that contacting sick-listed workers or helping them return to work is not a legal requirement; but rather a duty of care (though there is legislation covering protection after return to work). Most important is the focus on recovery of health at work).*

TABLE 3: WORK FOR SICK AND DISABLED PEOPLE

Authors	Key features (*Additional reviewers' comments in italics*)
Table 3b: Sickness absence and return to work *continued*	
(HSE 2005) Guidance	**Working together to prevent sickness absence becoming job loss: Practical advice for safety and other trade union representatives** (Health and Safety Executive, UK) Work is essential to health, well-being, and self-esteem. When ill health causes long-term sickness absence, a downward spiral of depression, social isolation, and delayed recovery may make returning to work difficult and less likely. Reducing long-term sickness absence helps maintain a healthy and productive business and safeguards everybody's jobs. Following illness, injury or the onset of disability:
	• Starting everyday activities again, like going to work, helps people to feel better;
	• Any remaining pain or discomfort can often be managed at work, with the right adjustments;
	• Work that is well managed is good for people's health, whilst staying off work can make them feel worse;
	• The barriers to returning to work often arise from personal, work or family-related problems, rather than the original health condition itself;
	• Early intervention by employers, working in partnership with safety and other trade union representatives, significantly increases the chances of people who are off sick returning to work.
	• Successful return to work depends on constructive cooperation between everyone involved.
(Talmage & Melhorn 2005) Guidance	**Physician's guide to return to work** (American Medical Association) Why should physicians encourage early and ultimate return to work whenever possible? 'Simply stated, **because it is usually in the patient's best interest to remain in the workforce**' (*their emphasis*). (*Elements of the physician guidance are in Tables 5 and 6*).

TABLE 3: WORK FOR SICK AND DISABLED PEOPLE

Authors	Key features (*Additional reviewers' comments in italics*)

Table 3b: Sickness absence and return to work *continued*

Authors	Key features
(Mowlam & Lewis 2005) DWP Research Report	**Exploring how general practitioners work with patients on sick leave** (Department for Work and Pensions, UK) Qualitative study of GPs' approaches to managing sickness absence and to assisting patients in returning to work. Dealing with sickness absence is a daily issue for GPs. Most absences are short; more problematic and sometimes lengthier absence is particularly associated with back pain, depression, stress and anxiety. GPs commented on the rising prevalence of absence due to workplace stress arising from poor relationships at work, and rising workloads and pressure. The view among GPs is that sickness absence is almost always genuine. However, patients' behaviour and motivation is also said to be influenced by issues such as subjective reactions to the experience of illness, organisational culture and financial circumstances. There is a widespread view among GPs that work can be of therapeutic benefit for a range of physical and psycho-social reasons. This view is qualified, however, where patients worked in low-paid jobs of low social status, and where the job itself caused or exacerbated a physical or psychological condition. These views are influenced by GPs' own personal views about the value of work, as well as observations of patients and research.

TABLE 3: WORK FOR SICK AND DISABLED PEOPLE

Authors	Key features (*Additional reviewers' comments in italics*)

Table 3b: Sickness absence and return to work *continued*

Authors	Key features
(Young *et al.* 2005) Narrative review	**A developmental conceptualization of return to work** There is general agreement among all stakeholders that a safe, timely, and sustainable return to productivity is desirable. The traditional medical model of return to work is inadequate because it assumes that capacity for work is primarily dependent on the nature and severity of the clinical condition. In reality, return to work is more complex because many health conditions are persistent or recurrent, and because personal and environmental factors are important. Return to work is not an isolated event but is better viewed as an evolving process that is influenced by different factors at different times. Phase 1 – Off Work. The worker is not necessarily totally incapacitated for work throughout this phase. This phase ends when a suitable return to work opportunity is available and the worker is about to attempt work re-entry. Phase 2 – Re-entry. During this phase the worker re-enters work either in his or her previous work(place) or at some alternative. Stakeholders go through a process of determining if and how work can be undertaken in a way that is satisfactory to all parties. This phase may include making work adjustments and concludes when the worker actually commences work. Phase 3 – Maintenance. During this phase the worker strives to meet work demands and monitoring of performance is likely. This is essentially a matter of demonstrating sustained employability, which may sometimes be open-ended. Phase 4: Advancement. Many sick or disabled people first enter low-status, low-paid or insecure work. Sustained employment may lead to raised 'human capital' and the potential for 'better' work. (*Young et al's original Phase 4 focused on qualifications and promotion, which has been adapted to the UK and social security context.*) This model of return to work has implications for rehabilitation interventions, outcome measurement, and research.

TABLE 3: WORK FOR SICK AND DISABLED PEOPLE

Authors	Key features (*Additional reviewers' comments in italics*)

Table 3b: Sickness absence and return to work *continued*

(ACOEM 2005) Report	**Preventing needless work disability by helping people stay employed** (American College of Occupational and Environmental Medicine) There is strong evidence that activity is necessary for optimal recovery from injury/illness/surgery, while inactivity delays it. Moreover, for an array of conditions including depression, chronic pain, fibromyalgia, and chronic fatigue syndrome, simple aerobic physical activity has been shown to be an effective treatment. There is also evidence that remaining at or promptly returning to some form of productive work improves clinical outcomes as compared to passive medical rehabilitation programs. Therefore, the ACOEM Practice Guidelines consistently recommend exercise, active self-care, and the earliest possible safe return to work. Better management of the stay at work and return to work process (including key non-medical aspects) will improve outcomes, support optimal health and function for more individuals, and encourage their continuing contribution to society.
(Franche *et al.* 2005) Systematic review	**Workplace-based return-to-work interventions: a systematic review of the quantitative literature.** 10 studies of workers compensation claimants were of sufficient quality to be included in the review. There was strong evidence that work disability duration is significantly reduced by work accommodation offers and contact between healthcare provider and workplace; and moderate evidence that it is reduced by interventions which include early contact with worker by workplace, ergonomic work site visits, and presence of a return-to-work coordinator. For these five intervention components, there was moderate evidence that they reduce costs associated with work disability duration. There was limited evidence on the sustainability of these effects. There was mixed evidence regarding any direct impact on quality-of-life outcomes. (Importantly, however, this review found no evidence that return to work had any adverse impact on quality of life) (*Also in Table 5*).

TABLE 3: WORK FOR SICK AND DISABLED PEOPLE

Authors	Key features (*Additional reviewers' comments in italics*)

Table 3b: Sickness absence and return to work *continued*

Authors	Key features
(FOM 2005) Handbook	**The Health and Work Handbook: Patient care and occupational health: a partnership guide for primary care and occupational health teams** (Faculty of Occupational Medicine/Royal College of General Practice/Society of Occupational Medicine, UK) Work is the best way to achieve economic independence, prosperity and personal fulfilment; it also helps reduce health and social inequalities. Helping patients to stay in work, or return to work, after absence due to illness or injury, is an important part of the therapeutic process, improves health outcomes in the long-term, is essential to restoring quality of life, and is an indicator of successful outcome of treatment. Conversely, long-term sickness absence can lead to job loss, long-term incapacity, social exclusion, and health inequalities. Primary care teams and occupational health professionals have a central role to play in return to work. GPs are well placed to offer simple fitness for work advice to their patients and to provide the focused support necessary to assist their recovery and retention in work. Occupational health professionals can provide more specialist advice and support in developing return to work programmes which will ensure that workers can return to work and that such return can be sustained. Close working and effective communication between primary care and occupational health professionals is essential.
(FPH/FOM 2006) Guidance	**Creating a healthy workplace** (Faculty of Public Health/Faculty of Occupational Medicine) The workplace has a powerful effect on the health of employees. How healthy a person feels affects his or her productivity, and how satisfied they are with their own health, both physical and psychological. When organisations proactively improve their working environments by organising work in ways that promote health, all the adverse health-related outcomes, including absence and injuries, decrease (making a strong business case for creating a healthy workplace). Health promotion initiatives will only be effective under conducive management conditions, primarily those that stimulate employee satisfaction. Other important factors include how work is organised carried out, physical working conditions, employee consultation/involvement, and organisational policies, procedures and rules. Guidance is provided on how to employers can achieve a healthy workplace – there is a focus on mental well-being/stress and musculoskeletal conditions: interventions include risk assessment and hazard control, smoke-free workplaces, promotion of physical activity and healthy eating, and sickness absence/rehabilitation policies.

TABLE 4: THE IMPACT OF WORK ON THE HEALTH OF PEOPLE WITH MENTAL HEALTH CONDITIONS

Authors	Key features (*Additional reviewers' comments in italics*)
Table 4a: Severe mental illness	
(Schneider 1998) Narrative review	**Work interventions in mental health care** Argued for a reconsideration of the role of work in psychiatric treatment and rehabilitation for people with severe mental disorders. Offered five perspectives on why psychiatry should place greater emphasis on work: • Ideological principles and social justice • Macro-economic considerations • Demand on the part of service users themselves • The changing context of mental health care and of public perspectives on mental illness • The evidence of clinical benefits from constructive occupation, which are relevant to evidence-based health care. Under the last heading, reviewed the historical evidence on constructive occupation in UK mental hospitals and in earlier UK vocational rehabilitation settings, though concedes much of that was observational and now out of date. Considered how work might improve mental health but also recognised that 'there is a possibility that for people with mental health problems work itself might be psychologically detrimental'. However, 'such reservations serve, not to rule out work, but to direct attention to the quality of employment opportunities available to people with mental health problems, to matching jobs to abilities, and to taking into account the conditions in which a person works'. Concluded that 'evidence of clinical benefits – – is on balance positive'.
(Baronet & Gerber 1998) Systematic review	**Efficacy of psychiatric rehabilitation – evidence on four models** Definition: a psychiatric rehabilitation programme is one whose primary focus is on improving clients' skills in order to minimise the impact of the psychiatric illness on their functional capacity. 22 studies on the Assertive Community Treatment Model showed positive effects on symptoms and health care use, but did not improve global functioning and had mixed effects on quality of life. Only 4 studies had occupational outcomes, with conflicting results. 25 studies of Case Management improved social and global functioning with mixed effects on quality of life. Case management programmes that emphasised use of vocational services had a positive impact on occupational status. 11 studies of Supported Employment produced positive effects on the ability to gain employment and on sustained employment over time. However, there was patient selection for entry to these programmes, including clinical stability and willingness to participate in vocational training and seeking employment. 5 studies of Educational Rehabilitation were, overall, associated with improved educational status which was generally followed by improved occupational status. Again, however, there was probably some selection of a higher functioning sub-group of patients. (*Reviewed studies published from 1987-1996*).

TABLE 4: THE IMPACT OF WORK ON THE HEALTH OF PEOPLE WITH MENTAL HEALTH CONDITIONS

Authors	Key features (*Additional reviewers' comments in italics*)

Table 4a: Severe mental illness *continued*

(Barton 1999) Meta-analysis	**Psychosocial rehabilitation services in community support systems**

The clinical characteristics and service needs of people with serious and persistent mental illness vary over the course of the illness and the individual's life. Long-term studies show that over a number of years, recovery is a realistic possibility and about 60% of patients achieve symptomatic remission and normal role functioning. Reviews of Supported Employment have consistently found positive employment outcomes but no evidence that employment gains are generalized to other outcomes (see Crowther et al 2001 below). Cost-effectiveness studies show an average reduction of >50% in health care costs, particularly due to reduced hospitalisation. (*No further data was presented on cause-effect relationships between work and symptoms*).

(Simon *et al.* 2001) Systematic review	**Depression and work productivity**

Major depression is one of the health conditions associated with the greatest work loss and work cutback. There is a strong dose-response relationship between the severity of depression and level of psychosocial disability. There is a strong association between improvement in depression and improved capacity for work, though that tends to lag behind symptomatic improvement and remains vulnerable to clinical relapse. (No further data was presented on cause-effect relationships between work and symptoms).

(Crowther *et al.* 2001b) (*Crowther et al. 2001a*) Cochrane Review	**Helping people with severe mental illness to obtain work**

(*Systematic review and Cochrane review including 18 randomised controlled trials (RCT) of reasonable quality*.) The main finding was that on the primary outcome measure (number in competitive employment) there is strong evidence that Supported Employment is effective and significantly more effective than Pre-vocational Training; for example at 18 months 34% of people from Supported Employment interventions were employed versus 12% from Pre-vocational Training (RR random effects (unemployment) 0.76 95% CI 0.64 to 0.89, NNT 4.5). Clients in Supported Employment also earned more and worked more hours per month than those in Pre-vocational Training.

Pre-vocational Training was no more effective than standard community care or hospital care in helping clients to obtain competitive employment (strong evidence).

Multiple but heterogeneous RCTs showed that neither Supported Employment nor Pre-vocational Training produced any significant effect on clinical outcomes such as severity of symptoms, hospitalisation rates for mental illness, self-esteem, or quality of life (*moderate evidence*). Data on health care costs were inconclusive. (*Importantly, however, the review did not describe any evidence that either Supported Employment or Pre-vocational Training caused any significant deterioration in the psychiatric condition*).

TABLE 4: THE IMPACT OF WORK ON THE HEALTH OF PEOPLE WITH MENTAL HEALTH CONDITIONS

Authors	Key features (*Additional reviewers' comments in italics*)
	Table 4a: Severe mental illness *continued*
(Schneider *et al.* 2002) (Scheider *et al.* 2003) Systematic review, expert consultation, UK policy paper	**Occupational interventions and outcomes for people with severe mental disorders** This review built on a Cochrane Review (Crowther et al 2001 - see above) and included 225 studies Studies of various forms of Sheltered Employment provided conflicting evidence on occupational outcomes. There was limited and conflicting evidence of any effect on quality of life, self-esteem and physical, mental and social functioning. Altogether, there was no conclusive evidence of effectiveness and some evidence Sheltered Employment might have detrimental effects. Studies of Training and Supported Education focused mainly on educational outcomes and did not report or provide any clear evidence on occupational outcomes. No evidence was reviewed on clinical outcomes. Found that the strongest evidence was for Supported Employment (largely based on the Cochrane Review by Crowther *et al* 2001). Found additional, limited evidence (Bond *et al.* 2001) that clients who did a substantial amount of competitive employment had greater satisfaction with vocational services, finances and leisure activities, and showed greater improvement in self-esteem and psychiatric symptoms. However, most of the evidence on Supported Employment is from the US, and these interventions need to be evaluated in a UK context. The authors were optimistic that: 'Social inclusion through employment is a more realistic prospect for people with (severe) mental health problems than ever before. There are six reasons why this is so in the UK today. Firstly, as always, there is a steady demand for paid work on the part of people with mental health problems. Secondly, there is broad legislative provision to protect the right to work of all disabled people. Thirdly, there are policy guidelines, with the ultimate objective of increasing social inclusion, that highlight the importance of employment. Fourthly, the benefits system is becoming progressively more flexible in relation to some forms of employment. Fifthly, there is a growing body of practice knowledge about how to help people with mental health problems achieve employment. And, finally, there is some sound evidence of the effectiveness of occupational interventions.'

TABLE 4: THE IMPACT OF WORK ON THE HEALTH OF PEOPLE WITH MENTAL HEALTH CONDITIONS

Authors	Key features (*Additional reviewers' comments in italics*)

Table 4a: Severe mental illness *continued*

(Twamley *et al.* 2003) Meta-analysis	**Vocational rehabilitation in schizophrenia and other psychotic disorders** Review of 11 RCTs, 9 of which were of Individual Placement and Support (IPS) or Supported Employment (SE). Meta-analysis showed that 51% of patients who received IPS/SE were in competitive employment, versus 18% of those who received conventional vocational rehabilitation (weighted mean effect size = 0.79). As part of the narrative background, stated that 'The multiple benefits of work rehabilitation and employment for patients with schizophrenia are well documented, and include increased income, external structure, achievement of a valued social responsibility, greater socialisation, more opportunities to use skills, and improved self-esteem' (and provided references to support these statements, but did not actually analyse these non-vocational data).
(Marwaha & Johnson 2004) Systematic Review	**Schizophrenia & Employment** Many people with schizophrenia say they want to work but there are wide variations in their reported employment rates. Most recent European studies report rates between 10 - 20%, while the rate in the US is less clear. The employment rate in schizophrenia appears to have declined over the last 50 years in the UK. There is a higher level of employment among first-episode patients. A large and consistent evidence base supports an association between good pre-morbid functioning and favourable employment outcome in schizophrenia. Barriers to getting employment include stigma, discrimination, fear of loss of benefits and a lack of appropriate professional help. The most consistent predictor of employment is previous work history. There is strong evidence from multiple studies of a correlation between working and positive outcomes in social functioning, symptom levels, quality of life and self esteem. However, few of these studies control for important baseline factors and a clear causal relationship has not been established.
(Bond 2004) Narrative review	**Supported employment for people with severe mental illness** Review of 4 studies of the conversion of day treatment to supported employment and 9 RCTs comparing supported employment to a variety of alternative treatments. These two lines of research suggest that 40-60% of patients enrolled in a supported employment programme obtain competitive employment compared with <20% of similar patients in the control groups. Enrolment in supported employment, by itself, has no systematic impact on non-vocational outcomes such as re-hospitalisation or quality of life. However, those patients who do hold competitive jobs for a sustained period of time show better symptom control and improved self-esteem. (*That last conclusion appears to be based on the single study by Bond et al 2001*).

TABLE 4: THE IMPACT OF WORK ON THE HEALTH OF PEOPLE WITH MENTAL HEALTH CONDITIONS

Authors	Key features (*Additional reviewers' comments in italics*)

Table 4b: Minor/moderate mental health problems

Authors	Key features
(Harnois & Gabriel 2000) WHO/ILO Report	**Mental health and work: impact, issues and good practices** There is growing awareness of the role of work in promoting or hindering mental well-being and its corollary – mental illness. Most mental health professionals agree that the workplace environment can have a significant impact on an individual's mental well-being (both positively and negatively). Mental health problems are the most common cause of illness, sickness, disability and loss of productivity. 15–30% of people will experience some form of mental health problem at some time in their lives. Even if work is not the primary cause of mental illness (*which is usually multifactorial – (Wessely 2004)*), mental illness impacts on work and is therefore an occupational health issue. Considers major myths about mental illness in the workplace and the evidence that they are untrue: • Myth 1: Mental illness is the same as mental retardation. • Myth 2: Recovery from mental illness is not possible. • Myth 3: Mentally ill employees tend to be second-rate workers (even after effective treatment). • Myth 4: People with psychiatric disabilities cannot tolerate stress on the job. • Myth 5: Mentally ill individuals are unpredictable, violent and dangerous (even after effective treatment). Argues that the workplace is an appropriate environment in which to educate individuals about, and raise their awareness of, mental health problems; to promote good mental health practices; for the recognition and early identification of mental health problems; and to establish links with local mental health services for referral, treatment and rehabilitation. Provides examples of good practice in mental health promotion in the workplace; management of workers who develop mental health problems; and vocational rehabilitation models/programmes for workers with long-term mental health problems. Mental health at work should encompass individual and organizational dimensions. (*This is a somewhat 'medical model' approach.*)
(Dunn *et al.* 2001) Systematic review	**Physical activity dose-response effects on outcomes of depression and anxiety** Observational cross-sectional studies demonstrate that greater amounts of occupational and leisure time physical activity are generally associated with less symptoms of depression. Quasi-experimental studies show that light, moderate and vigorous intensity exercise can reduce symptoms of depression. There is mixed evidence on the relationship between cardiorespiratory fitness and reduction in symptoms. There is no evidence on a dose-response relationship. There is limited evidence on anxiety.

TABLE 4: THE IMPACT OF WORK ON THE HEALTH OF PEOPLE WITH MENTAL HEALTH CONDITIONS

Authors	Key features (*Additional reviewers' comments in italics*)

Table 4b: Minor/moderate mental health problems *continued*

Authors	Key features
(Glozier 2002) Narrative review	**Mental ill health and fitness for work** The term 'common mental disorders' includes those who are considered as 'cases' by such measures as the General Health Questionnaire (GHQ), the most commonly used measure of 'stress' or 'mental ill health' in occupational studies, and those with the minor, and usually mixed, anxiety and depression often seen in primary care. At any one time some 20–35% of the working population fall into the former category and approximately 10% the latter. 'The relationship between mental ill health and fitness for work is still very unclear. Much of the research has looked at specific cohorts and often ignored factors external to the workplace. The advocates of the effects of "workplace stress" should not lose site of the fact that being unemployed is associated with twice the level of psychiatric morbidity of any employed group (teachers and nurses being possible exceptions). However, the working environment can be an important determinant of both mental ill health and, for many, well-being.'
(RCP 2002) Report	**Employment opportunities and psychiatric disability** • Work plays a central role in people's lives and is a key factor in social inclusion. • Work is important in maintaining and promoting mental and physical health and social functioning. Being in work creates a virtuous circle; being out of work creates a vicious circle. • Work is important in promoting the recovery of those who have experienced mental health problems. (*Largely based on the evidence about work and unemployment (as in Table 1) and the Cochrane Review by Crowther et al 2001 - see above. Does not document any evidence on the health impact of work for patients with mild/moderate mental health problems*).
(Thomas et al. 2002) Conceptual, narrative review	**Job retention & mental health** Reviews the literature on job retention and mental health problems, including Employee Assistance Programmes, the Social Process Model, the Case Management Approach, and the Avon & Wiltshire Job Retention Pilot. Concludes that: 'Work has a positive effect on people with mental health problems in relation to self-esteem, self-identity, providing scheduled activity and purpose, and symptom management. Conversely, losing a job or being demoted often has a negative effect on people with mental health problems' (*but does not provide references or linkage to the evidence for this conclusion*).

TABLE 4: THE IMPACT OF WORK ON THE HEALTH OF PEOPLE WITH MENTAL HEALTH CONDITIONS

Authors	Key features (*Additional reviewers' comments in italics*)
Table 4b: Minor/moderate mental health problems *continued*	
(Fryers *et al.* 2003) Systematic review	**Social inequalities and the common mental disorders: a systematic review of the evidence** The focus of this review was on less severe 'common mental disorders' - 'neurotic' conditions dominated by anxiety, depression or a combination of both – which contribute substantially to all morbidity. The review included 9 general population studies in developed countries (4 in UK) since 1980, with samples of >3,000 adults of working age, broad social class range and mental illness identified by validated instruments (5 General Health Questionnaire, 6 clinical diagnostic criteria). Eight out of nine studies provided evidence of a positive association between less privileged social position and higher prevalence of common mental disorders. The most consistent relationship was with unemployment, less education (OR 1.26–2.82), lower income, or lower material standard of living (OR 1.53–2.59); occupational-based social class was least consistent. Importantly, no study showed any negative association. The authors concluded that common mental disorders are significantly more frequent in socially disadvantaged populations.
(Michie & Williams 2003) 2003) Systematic review	**Reducing work related psychological ill health and sickness absence.** Review of evidence about the work factors associated with, and about successful interventions to prevent or reduce, psychological ill health and sickness absence. Key work factors associated with psychological ill health and sickness absence in staff include long hours worked, work overload and pressure, and the effects of these on personal lives; lack of control over work; lack of participation in decision making; poor social support; and unclear management and work role. (*This review apparently focused mainly on a particular set of 'psychosocial aspects of work'.*) There was some evidence that sickness absence was associated with poor management style. Successful interventions that improved psychological health and levels of sickness absence used training and organizational approaches to increase participation in decision making and problem solving, to increase support and feedback, and to improve communication. It is concluded that many of the work related variables associated with high levels of psychological ill health are potentially amenable to change, as shown in intervention studies that have successfully improved psychological health and reduced sickness absence.

TABLE 4: THE IMPACT OF WORK ON THE HEALTH OF PEOPLE WITH MENTAL HEALTH CONDITIONS

Authors	Key features (Additional reviewers' comments in italics)

Table 4b: Minor/moderate mental health problems continued

(Seymour & Grove 2005) BOHRF Report	**Workplace interventions for people with common mental health problems** (British Occupational Health Research Foundation). This report defined common mental health problems as those that: • occur most frequently and are more prevalent; • are mostly successfully treated in primary rather than secondary care settings; • are least disabling in terms of stigmatising attitudes and discriminatory behaviour. Concluded that for the: 1. Job retention of employees at risk: *** individual approaches to stress reduction, management and prevention for a range of health care professionals are effective and are preferable to multi-modal approaches. 2. Rehabilitation of employees with sickness absence associated with mental health problems: *** cognitive behavioural (CBT) interventions are effective and they are more effective than other intervention types. CBT is most effective for workers in high-control jobs. ** brief (up to 8 weeks) therapeutic interventions such as individual counselling are effective for employees with job-related or psychological distress Note that outcome measures were very mixed and often combined self-report and observational indices: however, 'making people better' is not the same as 'getting them back to work'. (*This review includes theoretical arguments why job retention is good for mental health, but does not provide any direct empirical evidence.*) Round Table Discussions at the launch of the Report supported these conclusions and expanded upon several points:
Discussions	1. Common health problems (CHP) at work are generally labelled 'stress' but the same problems outwith work tend to be given different diagnoses. 2. Health professionals need to take a more balanced approach to CHPs. E.g. GPs need to be informed that work can have positive therapeutic effects (evidence shows that work is more often good for mental health than bad for it). 3. Practical tools need to be developed to implement evidence-based individual level interventions. 4. Good quality research is needed into organisational level interventions. 5. Good cost-benefit studies are required to establish the business case and persuade employers and others to invest in better management of CHPs. 6. The question was raised whether the findings in this Report were compatible with the HSE Stress Management Standards. The response was that these were really looking at two different things. The stress management standards focus specifically on workplace issues and interventions, and acknowledge that there is currently a lack of evidence on some of these issues. Even if workplace issues were resolved, there would still be a lot of CHPs and these would impact on work. This Report focuses on their management

TABLE 4: THE IMPACT OF WORK ON THE HEALTH OF PEOPLE WITH MENTAL HEALTH CONDITIONS

Authors	Key features (*Additional reviewers' comments in italics*)

Table 4c-i: Stress: The impact of work on mental health

Authors	Key features
(Edwards & Cooper 1988) Narrative review	**The impact of positive psychological states on physical health: a theoretical framework, review of the evidence and methodological issues** While much research has focused on the impacts of negative psychological states, such as stress, on physical health, there has been relatively little research into the effects of positive psychological states. This paper proposes a theoretical framework in which stress is defined as 'a negative discrepancy between an individual's perceived state and desired state, provided that the presence of this discrepancy is considered important by the individual'. This model may be readily adapted to include positive emotional experiences by explicitly acknowledging positive discrepancies where the individual's perceived state meets or exceeds their desired state. This is analogous to Selye's discussion of 'eustress', which refers to the happy, healthy state of fulfillment. It is also consistent with positive psychological states such as job satisfaction, perceived quality of life and subjective well-being. Such positive psychological states may influence health by two main mechanisms: a) by evoking physiological responses that, in the long run, have direct beneficial effects on physical (*and mental*) health (this is the converse of the usual view of stress as a purely negative phenomenon); b) indirectly, by facilitating coping (with stress), leading to beneficial effects on physical (*and mental*) health. Thus, just as stress may have negative effects on physical (*and mental*) health, eustress may have positive or even curative effects on physical (*and mental*) health. Reviewed the evidence regarding the positive impact of eustress on health and concluded it was generally supportive, but various methodological problems prevented firm conclusions.

TABLE 4: THE IMPACT OF WORK ON THE HEALTH OF PEOPLE WITH MENTAL HEALTH CONDITIONS

Authors	Key features (*Additional reviewers' comments in italics*)

Table 4c-i: Stress: The impact of work on mental health *continued*

Authors	Key features
(Ursin 1997), (Eriksen & Ursin 1999), (Eriksen & Ursin 2002), (Eriksen & Ursin 2004) (Ursin & Eriksen 2004)	**Subjective health complaints (SHC)** General population surveys show that at least 75% of working-age adults report one or more bodily or mental symptoms in the past 30 days, the most common of which include tiredness (half), worry (a third) and depressed mood (a quarter). A third report three or more symptoms, and a quarter describe them as 'substantial'. The most common clusters are musculoskeletal, gastrointestinal or 'pseudoneurological' (e.g. fatigue, tiredness, vertigo or headaches), with overlap between them. Other studies show a high correlation between psychological, pain and musculoskeletal symptoms and high intra-individual variability over time, though initial reports tend to be higher (Steingrimsdottir et al. 2004). Such symptoms are often attributed to the 'stress' of modern life and work, but the prevalence is more or less equal across all societies studied (including UK), across time, and in primitive societies. Yet, despite these symptoms, about 80% of that population describe their health as 'good'. The vast majority are not receiving any health care and do not regard themselves as disabled with their symptoms. Ursin & Eriksen suggest that SHC are based on sensations from what most people regard as normal physiological processes. The level of complaints depends on combinations of high demands with low coping and high levels of helplessness and hopelessness. Increased SHC may be due to 'sensitisation' in which there is increased (possibly sustained) arousal and hyper-vigilance, with bodily sensations perceived as more severe and uncontrollable. A 'Cognitive Activation Theory of Stress' is hypothesised: there is no clear evidence to what extent this is a neurophysiological, neurobiological (endocrine, immunological or 'stress response'), psychological or behavioural phenomenon: it would seem to be primarily a matter of cognitive psychology for which there are plausible neurobiological mechanisms. When SHC become sufficiently intense and long-lasting they may reach the DSM-IV diagnostic criteria for somatisation disorder or undifferentiated somatoform disorder, but that requires a certain duration, a certain number of complaints, the condition causing psychological distress, and SHC representing a clinically significant impairment. However, the distribution of these complaints is continuous, with no clear threshold or cut-off for clinically significant vs. normal symptoms. The current social, insurance and health care problem is with SHC that do not reach these strict diagnostic criteria. The questions are why some people are more affected by such SHC, why some are less able to tolerate them, and why some seek health care or become incapacitated for work. Doctors and the general public have conceptual difficulties about disease/illness/sick certification for SHC, and are reluctant in principle to accept psychological and social problems as the basis for sick certification (Haldorsen et al. 1996). When tested with sample vignettes, however, doctors' decisions on sick certification are more or less random and in practice, patients regularly seek and doctors regularly issue sick certificates for SHC. (*continued*)

TABLE 4: THE IMPACT OF WORK ON THE HEALTH OF PEOPLE WITH MENTAL HEALTH CONDITIONS

Authors	Key features (*Additional reviewers' comments in italics*)

Table 4c-i: Stress: The impact of work on mental health *continued*

Authors	Key features (*Additional reviewers' comments in italics*)
(Ursin 1997), (Eriksen & Ursin 1999), (Eriksen & Ursin 2002), (Eriksen & Ursin 2004) (Ursin & Eriksen 2004)	**Subjective health complaints (SHC)** (continued) (Ihlebaek & Eriksen 2003) found no significant difference in the prevalence of SHC between major occupational groups. However, the 10% of workers who account for 82% of sickness absence have a significantly higher level of SHC (P<.001) (Tveito *et al.* 2002). Analysis of a sample of the Norwegian general population (n=1014) shows no significant difference in the prevalence of SHC in the unemployed though those who are retired/disability pension do show a significantly higher level of SHC (C Ihlebaek personal communication).
(van der Doef & Maes 1999) Systematic review	**The Job Demand-Control-(Support) Model and psychological well-being** Review of 63 studies published 1979-1997. The literature gives considerable support for the strain hypothesis - workers with high demands and low control experience the lowest levels of psychological well-being. 28/41 studies support an association with general psychological well-being, 18/30 with job satisfaction and 7/8 with job-related psychological well-being. However, only two of nine longitudinal studies showed an additive effect of demands and controls. 15 of 31 studies provided (partial) support for the hypothesis that control has a buffering effect between demands and well-being, but again mainly in cross-sectional studies. Most of the positive evidence is in men. There is conflicting evidence for the moderating influence of job control or social support. Only aspects of job control directly relevant to the specific demands of the specific job appear to moderate the impact of high demands on well-being. Furthermore, certain sub-populations appear to be more vulnerable to strain while others benefit more from high control. (*Using the present review's evidence rating system, the studies reviewed by van der Doef & Maes provide mixed or conflicting evidence that there is an association between demands and control and psychological well-being. No data is presented on the strength of this association. As these studies are mainly cross-sectional, this evidence does not establish cause and effect.*)
(Viswesvaran *et al.* 1999) Meta-analysis	**The role of social support in work stress** Social support was defined as 'the availability of helping relationships and the quality of those relationships. Highly complex meta-analytical modeling of 68 studies suggested that social support acts in a threefold manner. Its primary role is to reduce strains (estimated true correlation across categories = - 0.21). Its secondary roles are to reduce the effects of the stressors themselves (estimated true correlation = -0.12) and to alleviate the effects of stressors on strains (cumulated R^2 = - 0.03). However, all of these figures imply that only a small fraction of the variance is explained i.e. social support only has a weak effect on work stress (*strong evidence*). Further analysis showed that social support does not appear to function as a mediator or as a suppressor variable in the stressor–strain relationship. Neither does it seem likely that support is mobilized primarily in the presence of stressors.

TABLE 4: THE IMPACT OF WORK ON THE HEALTH OF PEOPLE WITH MENTAL HEALTH CONDITIONS

Authors	Key features (*Additional reviewers' comments in italics*)

Table 4c-i: Stress: The impact of work on mental health *continued*

Authors	Key features
(Salovey *et al.* 2000) Narrative review	**Emotional states and physical health** The association between emotional states and increased reports of physical complaints is well established: physical illness and pain can cause anxiety or depressed mood; the premise that emotional arousal can cause changes in physical health is the focus of this review. Considers the evidence for several possible mechanisms: 1. Positive and negative emotions can have direct effects on physiology, especially the immune system. 2. Emotions influence people's perceptions, interpretation and reporting of their bodily status, sensations and health. 3. Positive emotions provide greater psychological resources that enable people to cope more effectively with health problems. 4. Positive and negative emotions influence people adopting healthy or unhealthy behaviours (e.g. eating, smoking, alcohol or exercise) and seeking health care. 5. Emotional states influence social relationships and whether or not people receive social support from others. Concludes that, by all of these mechanisms, emotions may influence physical (*and mental*) health and well-being, either positively or negatively.
(Rhoades & Eisenberger 2002) Meta-analysis	**Perceived organizational support** Review of 73 studies with effect sizes expressed as product-moment correlation coefficients (r). Perceived organisational support (POS) was defined as employees' general belief that their work organisation valued their contribution and cared about their well-being. Meta-analysis showed that POS was associated with three major categories of beneficial treatment: 1) fairness, 2) supervisor support and 3) organisational rewards and favourable job conditions. POS was related to favourable outcomes for employees e.g. job satisfaction ($r=0.62$) positive mood ($r=0.49$) and the organisation e.g. organisational commitment ($r=0.67$), performance ($r=0.20$), desire to remain with the organisation ($r=0.66$) and turnover intention ($r=-0.51$). These relationships depended on employees' belief that the organisation's actions were discretionary, feeling of obligation to aid the organisation, fulfilment of socio-emotional needs, and performance-reward expectancies, which is all consistent with organisational support theory.

TABLE 4: THE IMPACT OF WORK ON THE HEALTH OF PEOPLE WITH MENTAL HEALTH CONDITIONS

Authors	Key features (*Additional reviewers' comments in italics*)
Table 4c-i: Stress: The impact of work on mental health *continued*	
(De Dreu & Weingart 2003) Meta-analysis	**Task and relationship conflict, team performance and team member satisfaction** Review of 30 studies with effect sizes expressed as corrected correlations. There were significant negative associations (*r*) between relationship conflict and team performance (- 0.25) and team member satisfaction (- 0.56); and between task conflict and team performance (- 0.20) and team member satisfaction (- 0.32).
(de Lange *et al.* 2003) Systematic review	**The demand-control-support model** Review of 45 longitudinal studies, of which 19 were considered to be of sufficient quality on all methodological criteria. 16 of these 19 studies provided evidence of a causal relationship between one or more psychosocial aspects of work (demand, control or support) and self-reported measures of health or well-being (mainly psychological distress) over time. Eight of these 19 studies supported the strain hypothesis, though that effect was usually additive rather than multiplicative. *(Using the present review's evidence rating system, the studies reviewed by De Lange et al provide strong evidence of a causal relationship between job demands, control or support and self-reported psychological well-being, with mixed or conflicting evidence that these effects were additive. No data was presented on effect sizes, but see Faragher 2005 below).*
(Tsutsumi & Kawakami 2004) Systematic review	**The Effort-Reward Imbalance (ERI) model** Review of 45 studies. Twelve cross-sectional studies and three cohort studies all showed some significant associations (*P* <0.05, *but without Bonferroni adjustment for multiple comparisons*) between ERI and various psychological and psychosomatic symptoms. Odds Ratios in the cross-sectional studies generally ranged from 1.5 – 7.0 but in the cohort studies were only 1.4 – 2.6 (*there was no meta-analysis to combine these results*) *(Also in Table 6)*

TABLE 4: THE IMPACT OF WORK ON THE HEALTH OF PEOPLE WITH MENTAL HEALTH CONDITIONS

Authors	Key features (Additional reviewers' comments in italics)
Table 4c-i: Stress: The impact of work on mental health *continued*	
(van Vegchel *et al.* 2005) Systematic review	**The Effort-Reward Imbalance (ERI) model** Review of 45 studies, 23 of which were of psychosomatic symptoms or job-related well-being. Fourteen out of 16 studies found that various psychosomatic symptoms were associated with high effort-low reward or with over-commitment, with Odds Ratios ranging from 1.44 – 18.55 (*though 11 of these 14 positive studies were cross-sectional and there was no meta-analysis to combine these results*). All 7 studies of 'job-related well-being' found a strong association between ERI and measures such as 'burnout', job satisfaction and work motivation (*though all 7 were cross-sectional studies and there is probable confounding in these measures*) (*Also in Table 6*)
(Faragher *et al.* 2005) Systematic review and meta-analysis	**The relationship between job satisfaction and health** Job satisfaction shows the strongest relationship to employee health of any psychosocial characteristic of work that these authors evaluated. Job satisfaction is about the positive emotional reactions and attitudes an individual has towards their job. This meta-analysis found that job satisfaction correlated most strongly with mental health: self-reported 'burn-out' (corrected $r = 0.42$), self-esteem ($r = 0.48$), depression ($r = 0.43$) and anxiety ($r = 0.42$). (*That is equivalent to about 15% of variance in common. However, any causal interpretation is limited by the cross-sectional nature of most studies, and the likely confounding between subjective self-reports of 'job satisfaction' - which is by definition an emotional construct - and subjective self-reports of 'mental well-being' - which is again a matter of emotions*). The correlation with subjective 'physical illness' was more modest ($r = 0.29$).

TABLE 4: THE IMPACT OF WORK ON THE HEALTH OF PEOPLE WITH MENTAL HEALTH CONDITIONS

Authors	Key features (*Additional reviewers' comments in italics*)

Table 4c-i: Stress: The impact of work on mental health *continued*

Authors	Key features
(Bartley *et al.* 2005) Evidence review	**Work, non-work, job satisfaction and psychological health** (Health Development Agency, UK) There is now widespread recognition that the relationship between work and health goes well beyond specific occupational diseases and accidents to a broader relationship which may be understood in terms of three mechanisms: • Work which provides fulfillment and allows individuals control over their working lives confers considerable health benefit. • Jobs which are lacking in self-direction and control seem to confer far fewer health benefits, and people with such jobs seem to experience consistently higher rates of mortality and morbidity. • Absence of work in the form of unemployment produces considerable negative health effects. However, the precise ways in which these mechanisms work is subject to considerable debate. (*Provides very little evidence for these conclusions.*) British Household Panel Surveys show a decline in mean levels of job satisfaction in UK between 1991-2002, but any change is marginal and there is some conflict within this data e.g. both the numbers who are highly satisfied and the numbers with very low satisfaction decreased; the decrease was greatest in intermediate non-manual jobs and least in those in the poorest jobs with least control and most insecurity. (*General discussion about the relationship between trends in job satisfaction, psychological health, economic inactivity, and Incapacity Benefit (IB) trends in the context of social determinants of health*).

TABLE 4: THE IMPACT OF WORK ON THE HEALTH OF PEOPLE WITH MENTAL HEALTH CONDITIONS

Authors	Key features (*Additional reviewers' comments in italics*)

Table 4-c-i: Stress: The impact of work on mental health *continued*

(Bond *et al.* 2006) HSE Research Report Meta-analysis	**A business case for the Management Standards for stress** (*This title is misleading*) The most convincing evidence is for control (the extent to which people have a 'say' in the way they do their work). Nineteen longitudinal studies and laboratory experiments consistently showed that higher levels of control led to better business outcomes. Meta-analysis of 8 studies showed significant improvement in objective performance (effect size 0.23), performance rating (0.32), absenteeism (- 0.11) and turnover intention (- 0.21). Four out of five rigorous studies of organizational interventions clearly demonstrated that increasing job control improved absenteeism, turnover and performance (objectively measured and as rated by others), which translated into meaningful financial savings. There was limited evidence (from 1-3 studies each) that higher levels of **support** (the encouragement, sponsorship, and resources provided by the organisation, line management, and colleagues) lead to comparable improvements in these outcomes. Meta-analysis of 24 studies showed that poor work **relationships** have their greatest effect by reducing team performance (effect size – 0.22). The business case appeared weakest for **demands** (aspects of work to which people have to respond, such as work load, work patterns, and the work environment). High demands only have meaningful and consistent deleterious effects on business outcomes in laboratory experiments. In actual work organisations, high demands are not a good predictor of any business outcome, except when they are accompanied by lower levels of control.

TABLE 4: THE IMPACT OF WORK ON THE HEALTH OF PEOPLE WITH MENTAL HEALTH CONDITIONS

Authors	Key features (*Additional reviewers' comments in italics*)

Table 4c-ii: Stress: Management

| (Cox 1993) Conceptual narrative review | **Stress research and stress management**

 Early research report for the Health & Safety Executive that attempted to provide an overview, within the conceptual framework implied by current health and safety legislation, of the literature on work stress and stress management.

 'There is growing consensus on the definition of stress as a psychological state with cognitive and emotional components and on its effects on the health of both individual employees and their organisations' and 'there are now theories which can be used to relate the experience and effects of work stress to exposure to work hazards and to the harmful effects on health that such exposure might cause' (*though acknowledged the subjective and cross-sectional nature of much of that evidence*). Acknowledged that there was limited evidence at that time on the effectiveness of stress management programmes.

 Advocated a transactional model of stress that involved the Person in their Environment, and included a sequence of events starting with (work-related) demands or hazards, involving a set of evaluations and responses, and producing a psychological state that affects the health and well-being of the individual. Argued that work-related stress should be addressed by a health and safety approach to risk assessment and control, just as required for physical hazards. |

TABLE 4: THE IMPACT OF WORK ON THE HEALTH OF PEOPLE WITH MENTAL HEALTH CONDITIONS

Authors	Key features (*Additional reviewers' comments in italics*)

Table 4c-ii: Stress: Management *continued*

(Briner & Reynolds 1999) Conceptual review	**The costs, benefits and limitations of organizational level stress interventions**

Questioned the rationale and evidence that organizational level interventions which aim to reduce job stressors will reduce or eliminate stress responses in workers. The arguments for these interventions are: 1) tackling the presumed occupational causes of stress will be more effective than treating the individual effects (prevention is better than cure); 2) studies of individual level stress management interventions have shown only limited and short-term effectiveness, and hence organizational level interventions will be more effective (which is a *non sequitur*); 3) targeting interventions at the individual level is viewed as somehow 'blaming the victim' (which is illogical, c.f. health care). Argument 1) is most compelling, but that depends on a) organisational stress being the cause of adverse employee states and behaviours, and b) organisational interventions reducing these undesirable employee states and behaviours, and these effects being uniformly positive. The rubric 'stress' includes a complex set of job characteristics, employee states and behaviours for which there is unlikely to be a simple explanation or solution; there is no good explanation of the mechanisms (why or how) by which these job characteristics might cause these employee states and behaviours; although there is a lot of evidence of an association between individual 'stress' and job 'stressors' (i.e. we do not like how we respond to what we do not like), profound methodological weaknesses (cross-sectional studies and self-report) means that there is an almost complete absence of good evidence to establish a causal link. b) if the causal link is not established or weak, then organisation interventions will not be (very) effective; it is unclear why or how they will be effective; they may therefore have mixed or adverse effects; at the time of this thee was limited and mixed evidence that organisational interventions were effective. Briner & Reynolds summarise the evidence: 'if half a dozen variables measuring undesirable employee states and behaviours are measured over time following an intervention, some will change in a positive direction, some will change in a negative direction, and some will not change at all. In addition, if the same intervention is introduced in two sites or organisations the impacts and outcomes of that intervention will probably differ. Such findings do not suggest to us that these interventions work. Rather, they suggest that their effects are mixed, but broadly neutral.' They conclude on both rational and empirical grounds, and whatever the moral or social arguments, organisational level interventions are unlikely to be the panacea for the many undesirable employee states and behaviours that are attributed to 'stress.'

TABLE 4: THE IMPACT OF WORK ON THE HEALTH OF PEOPLE WITH MENTAL HEALTH CONDITIONS

Authors | **Key features** (*Additional reviewers' comments in italics*)

Table 4c-ii: Stress: Management *continued*

(Cox *et al.* 2000b)
Report

Research on work-related stress
(European Agency for Safety and Health at Work)

Stress at work is a priority issue of the European Agency for Safety and Health at Work.

There are essentially three approaches to defining stress: 1) an aversive or noxious characteristic of the work environment that may cause ill health (the 'engineering approach'); 2) the common physiological effects of a wide range of aversive or noxious stimuli (the 'physiological approach'); Both these approaches treat the worker as a passive vehicle whereby environmental stimuli produce physiological or psychological responses; they assume a simple stimulus-response paradigm and fail to allow for perceptual and cognitive processes and contextual modifiers. 3) conceptualises work stress in terms of the dynamic interaction between the worker and their work environment (the 'psychological approach'). There is now a consensus developing around this third approach, defining stress as a negative psychological state, with cognitive and emotional components, which is both part of and reflects a wider process of interaction between the person and their (work) environment. (*This combines two definitions given in the Report*).

Psychosocial hazards are defined as 'those aspects of work design and the organisation and management of work, and their social and environmental contexts, which have the potential for causing psychological, social or physical harm'.

Coping is an important part of the overall stress process. There is considerable individual variation in the experience of stress and in how and how well it is coped with.

Taking the psychological approach and in view of individual differences, measurement of stress must be based primarily on self-report of individual experience. The literature is divided on the extent to which 'negative affectivity' (*which Cox et al describe as a 'personality trait, though it may be better considered as negative affect, psychological distress or minor psychological morbidity, from whatever cause*) might influence such self-report. Cox et al suggest this might be overcome by triangulation of different kinds of evidence:

1. objective and subjective antecedents of the person's experience of stress;
2. their self-report of stress;
3. any changes in their behaviour, physiology or health status.

TABLE 4: THE IMPACT OF WORK ON THE HEALTH OF PEOPLE WITH MENTAL HEALTH CONDITIONS

Authors	Key features *(Additional reviewers' comments in italics)*

Table 4c-ii: Stress: Management *continued*

(Cox et al. 2000a) HSE Research Report	**Organisational interventions for work stress: a risk management approach.** (Overview of research and development work conducted by the Institute of Work, Health and Organisations, University of Nottingham Business School for the Health & Safety Executive.) Describes the origins and logical basis of a risk management approach to the reduction of work stress, and the strategy that frames its processes and procedures. The basic health and safety equation of Hazard–Risk–Harm is proposed as the conceptual framework for understanding the nature and management of work stress (Cox 1993 – above). The experience of stress is then the link between employees' exposure to the hazards of work and any subsequent and related harm to their health, in a direct causal chain. Defines psychosocial and organisational hazards as 'those aspects of work design and the organisation and management of work, and their social and environmental context that have the potential for causing psychological, social or physical harm'. Defines stress as 'an emotional experience that is complex, distressing and disruptive'; stress can arise from two different sources at work: a) anxiety about exposure, or the threat of exposure, to the more tangible and physical hazards of work, or b) exposure to problems in the psychosocial environment and with their social and organisational settings i.e. psychosocial and organisational hazards. Stress can be dealt with primarily at source, by reducing exposure to hazards that are regarded as stressful, or at the individual level by treating the experience of stress itself or its health effects. This report is primarily concerned with the former strategy; though the case examples used demonstrate that most interventions involve both organisational and individually-focused elements. Argues that the priority is prevention but in practice control strategies tend necessarily to be a mixture of approaches. States that existing research into the nature and effects of work stress is neither appropriate nor adequate as an assessment of the associated risks. Lays out the logic of a risk assessment strategy framed by current thinking in health and safety management: **Hazard identification:** reliable identification of stressors and degree of exposure for specific groups of employees. Argues that 'since many of the exposures that give rise to the experience of stress at work are chronic in nature, the proportion of employees that reporting a particular aspect of work as stressful may be a "good enough" exposure statistic'. **Assessment of harm:** Evidence that exposure to these hazards is associated with impaired (sic) health, in a wide range of health-related outcomes. *(continued)*

TABLE 4: THE IMPACT OF WORK ON THE HEALTH OF PEOPLE WITH MENTAL HEALTH CONDITIONS

Authors	Key features (Additional reviewers' comments in italics)
Table 4c-ii: Stress: Management continued	
(Cox *et al.* 2000a) HSE Research Report	**Organisational interventions for work stress: a risk management approach.** (*continued*) (Overview of research and development work conducted by the Institute of Work, Health and Organisations, University of Nottingham Business School for the Health & Safety Executive.) **Identification of likely risks:** explore the association between exposure to hazards and likely risks, to make some estimate of the size and/or significance of the likely risks. (*Acknowledges later that this is based on associations between stressors and health outcomes but argues that is "good enough evidence" for the overall risk management process*). **Description of the underlying mechanisms:** understand and describe the possible mechanisms by which hazards are associated with (sic) health related harms'. Admits that all of these steps are 'challenging' and largely based on employees' subjective perceptions but argues this constitutes 'employees' expert knowledge of work', their experience of work and of stressors, and can be cross-checked by psychometric properties, face, conceptual and concurrent validity, and consensus. Concedes that 'the scientific literature on risk reduction in relation to work stress is sparse.' (*The remainder of the Report is concerned with the practicalities of the risk management approach and with case studies*).
(Briner 2000) Conceptual review	**Relationships between work environments, psychological environments and psychological well-being.** Work environments have many characteristics that may affect both physical and mental well-being. The psychological work environment is the set of characteristics that affect how workers feel, think, and behave. For at least the last 30 years, both empirical research and theory concerning work and well-being have focused on the negative impacts of work and particularly on the impact of work-related stress on mental health. But there is much evidence that work can also have beneficial effects on health and that being in employment is generally less harmful than being unemployed.

TABLE 4: THE IMPACT OF WORK ON THE HEALTH OF PEOPLE WITH MENTAL HEALTH CONDITIONS

Authors	Key features (*Additional reviewers' comments in italics*)

Table 4c-ii: Stress: Management *continued*

(Rick & Briner 2000) Conceptual review	**Psychosocial risk assessment: problems and prospects.**

Considers current approaches to risk assessment for psychosocial conditions and some of the associated problems. Employers have a general 'duty of care' and statutory duties laid down in Health and Safety legislation and regulations apply equally to the physical and mental health and well-being of employees. These statutory duties include an explicit requirement to conduct risk assessments, to identify hazards to health and to reduce the risks as far as reasonably possible. That approach was designed primarily for physical hazards but it has been argued it should apply equally to psychosocial hazards (Cox *et al* 1993 - above).

However, Rick & Briner argued that there are a number of problems to that approach that raise questions about its validity:

- Physical hazards are usually specific and clear-cut, once they have been recognised. There are problems to identifying, defining or measuring specific psychosocial stressors.

- Physical hazards always have negative effects, at least above a certain threshold. Psychosocial stressors may have beneficial or harmful effects on health, and it is often not possible to define the point at which they become hazardous.

- Physical hazards usually have immediate effects. Any impact on health from psychosocial hazards may remain latent for months or years.

- Physical hazards are usually intrinsically harmful, even if their impact may depend to some extent on individual susceptibility. Whether or not work characteristics are psychosocial hazards is determined wholly or partly by the way the worker perceives them.

- Most physical hazards have a clear link to specific physical injury or disease. It is often difficult to establish any clear link or to prove a causal relationship between psychosocial hazards and mental health problems. Such mental health problems are often non-specific, they are commonly multifactorial in origin, there is no clear threshold when psychosomatic symptoms become illness, and it is often difficult to decide when they constitute 'harm'.

- Most physical hazards and harms are objective. Both psychosocial hazards and harms are subjective, based on self-reports, and open to all the potential psychosocial influences on that.

- In summary, there is considerable uncertainty about psychosocial hazards, about psychosocial harms, and about the relationship between them (Mackay *et al* 2004 – see below). (*continued*) |

TABLE 4: THE IMPACT OF WORK ON THE HEALTH OF PEOPLE WITH MENTAL HEALTH CONDITIONS

Authors	Key features (*Additional reviewers' comments in italics*)
Table 4c-ii: Stress: Management *continued*	
(Rick & Briner 2000) Conceptual review	**Psychosocial risk assessment: problems and prospects.** (*continued*) These fundamental and inherent limitations make it difficult to identify and assess potential psychosocial hazards in the workplace, to decide if they might cause harm, and to assess the risk that harm will develop in any individual worker. They also raise major questions about (diagnosis and) the likely effectiveness of any intervention. *(Similarly, European Commission's Directive on 'Measures to Encourage Improvements in the Safety and Health of Workers at Work 1989' states that employers should develop 'a coherent overall prevention policy which covers technology, organisation of work, working conditions, social relationships and the influence of factors related to the working environment' (Article 6.2))*
(Merz *et al.* 2001) Narrative review	**Disability and job stress: implications for vocational rehabilitation planning** Workers with disabilities, especially those with mental health problems, are at greater risk of stress-related disorders. First, the labour market may place them in 'poorer', more stressful and less satisfying jobs. Second, the health condition itself may make them particularly vulnerable to stress and less able to cope. Some health conditions, and particularly mental illness, can be exacerbated by unmanaged job stress. Third, some people with disabilities do not have the same social supports to help them cope with work-related stress. (*No evidence is presented to support these arguments*).
(Rick *et al.* 2001) HSE Research Report	**A critical review of psychosocial hazard measures** The term 'psychosocial hazards' was used to refer to work characteristics which could equally be termed 'stressors' or 'sources of stress'. Although many measures of psychosocial hazards were available, there was limited scientific evidence on their psychometric properties or reliability. Most of these were subjective self-reports and their validity was, at best, moderate. Most striking was the lack of any evidence on the predictive value of any of these 'hazards' for important (health) outcomes. Similar comments could be made about their utility.

TABLE 4: THE IMPACT OF WORK ON THE HEALTH OF PEOPLE WITH MENTAL HEALTH CONDITIONS

Table 4c-ii: Stress: Management *continued*

Authors	Key features (*Additional reviewers' comments in italics*)
(Rick *et al.* 2002) HSE Research Report	**Scientific knowledge to underpin standards for key work-related stressors** Reviewed the evidence on four questions: 1. Exposure to nine key stressors: The available evidence is almost entirely self-reported and cross-sectional; it is about how individuals perceive and experience their workplace, rather than about actual hazards or whether they are harmful. 2. Effects of these stressors: High workload, shift work, high job demands, low control, low support and bullying/harassment have a negative impact on health. In general, effect sizes are small to moderate. 3. Mechanisms through which stressors can affect health must allow for: a) the combined effects of multiple stressors b) differences between individuals c) indirect links between stressors and their outcomes, and d) non-linear relationships. 4. Organisational interventions: Socio-technical interventions to reduce workload, improve work scheduling or improve work organization are particularly effective at improving well-being and performance measures. Psychosocial interventions to improve decision authority latitude have similar effects. Other psychosocial interventions have lesser and inconsistent results. *Rick et al concluded that it is not possible to draw any firm conclusions 'about which stressors are most harmful, at what threshold they become harmful, how they operate, or what can be done to reduce their levels'.*
(Mackay *et al.* 2004) Scientific and conceptual review	**Management Standards and work-related stress in the UK: policy background and science** Review of the development of UK policy since the late 1990s, to support HSE 2004 below. The standards-based approach is based on the assumption that stress at work is a hazard with adverse health consequences, and hence should be addressed by the existing health and safety (HSE) framework for controlling risk. The challenge for the Health and Safety Commission was to devise an effective programme to reduce work-related stress. This started from a stress model that defined stress as an (individual) psychological response to (work) demands, and drew on literature and workshops with key stakeholders to develop a taxonomy of stressor areas: • Demands– including such issues as workload, work patterns and the working environment. • Control – how much say the person has in the way they do their work. (*continued*)

TABLE 4: THE IMPACT OF WORK ON THE HEALTH OF PEOPLE WITH MENTAL HEALTH CONDITIONS

Authors	Key features (*Additional reviewers' comments in italics*)
	Table 4c-ii: Stress: Management *continued*
(Mackay *et al.* 2004) Scientific and conceptual review	**Management Standards and work-related stress in the UK: policy background and science** (*continued*)
	• Support – which includes the encouragement, sponsorship and resources provided by the organisation, line management and colleagues
	• Relationships at work – which includes promoting positive working practices to avoid conflict and dealing with unacceptable behaviour.
	• Role – whether people understand their role within the organisation and whether the organisation ensures that the person does not have conflicting roles.
	• Change – how organisational change (large or small) is managed and communicated in the organisation,
	• Culture – the way in which organisations demonstrate commitment and have procedures that are fair and open.
	The approach to controlling these hazards was then to: 1. look for the hazards; 2. decide who might be harmed and how; 3. evaluate the risks and decide whether the existing precautions are adequate; 4. record your findings; and 5. review your assessment and revise if necessary. (*Which is the standard HSE approach to assessing and controlling risk.*)
	The authors acknowledged that there was strong evidence for an association between these work hazards and ill-health, but that there were well-documented methodological problems that made it difficult to demonstrate a causal relationship. They suggested that a causal link was supported by two broad lines of evidence:
	1) Biological mechanisms that might mediate the pathways between stress and various disease states:
	• homeostatic and allostatic changes in response to stress
	• neuroendocrine changes and alterations of autonomic function.
	• development of the metabolic syndrome and insulin resistance
	• inflammatory and immune responses which mediate the susceptibility to infection
	• psychological mechanisms such as anxiety, hypervigilance and risk taking
	(*Mackay et al provided references to literature in each of these areas, but this was essentially evidence to support these scientific theories, rather than evidence of the occurrence, prevalence or importance of these mechanisms in the context of work or work-related stress disorders*). (*continued*)

TABLE 4: THE IMPACT OF WORK ON THE HEALTH OF PEOPLE WITH MENTAL HEALTH CONDITIONS

Authors | **Key features** (*Additional reviewers' comments in italics*)

Table 4c-ii: Stress: Management *continued*

Authors	Key features
(Mackay *et al.* 2004) Scientific and conceptual review	**Management Standards and work-related stress in the UK: policy background and science** (continued) 2) Epidemiological and psychosocial evidence: This was essentially a brief review of the evidence on the Demand-Control model, social support and various additional psychosocial aspects of work. The authors acknowledged that much of this evidence was cross-sectional in nature and that 'there is insufficient evidence to answer these questions (about the relation between these stressors and health outcomes) with complete satisfaction'. (*See (Rick et al. 2001; Rick et al. 2002; de Lange et al. 2003)*). In light of that evidence, the authors considered the validity of the risk assessment approach for work-related stress. They acknowledged questions about whether risk assessment and risk management as applied to physical hazards was appropriate, because of the essentially psychological nature of the stress process and uncertainty about the relationship between hazard and harm. However, they emphasised the importance of and need to adhere to this fundamental European approach to health and safety. They then argued this approach was applicable to stress by: 1) accepting that subjective perceptions reflect and provide some measure of more 'objective' characteristics of the working environment 2) distinguishing current or reported levels of particular environmental features from some preferred or desired 'state to be achieved' and focusing on the latter, and 3) using a 'bottom-up approach' that takes account of workers' and employers' views and concerns and the local context.

TABLE 4: THE IMPACT OF WORK ON THE HEALTH OF PEOPLE WITH MENTAL HEALTH CONDITIONS

Authors	Key features (*Additional reviewers' comments in italics*)

Table 4c-ii: Stress: Management *continued*

(HSE 2004a)
UK Management
Standards
	Management Standards for tackling work-related stress
	Health & Safety Executive analysis tool covering six standard areas:
	Demands – includes issues like workload, work patterns, and the work environment.
	  Standard: employees indicate that they are able to cope with the demands of their jobs.
	Control – How much say the person has in the way they do their work.
	  Standard: employees indicate that they are able to have a say about the way they do their work.
	Support – Includes the encouragement, sponsorship and resources provided by the organisation, line management and colleagues.
	  Standard: employees indicate that they receive adequate information and support from their colleagues and superiors.
	Relationship – includes promoting positive working practices to avoid conflict and dealing with unacceptable behaviour.
	  Standard: employees indicate that they are nor subjected to unacceptable behaviours, e.g. bullying at work.
	Role – Whether people understand their role within the organisation and whether the organisation ensures that the person does not have conflicting roles.
	  Standard: employees indicate that they understand their role and responsibilities.
	Change – How organisational change (large or small) is managed and communicated in the organisation,
	  Standard: employees indicate that the organisation engages them frequently when undergoing an organisational change.
	And, in each area, - Systems are in place locally to respond to any individual concerns.

TABLE 4: THE IMPACT OF WORK ON THE HEALTH OF PEOPLE WITH MENTAL HEALTH CONDITIONS

Authors	Key features (*Additional reviewers' comments in italics*)

Table 4c-ii: Stress: Management *continued*

(HSE/HSL 2005) Workshop materials (Bowen *et al.* 2005) Workshop report	**Review of the risk prevention approach to occupational health: applying health models to 21st century occupational health needs** (UK Health & Safety Executive and Health & Safety Laboratory) Occupational health defined as: 'a state of physical, mental, and social well-being at work, and not merely the absence of disease and disability, that is influenced by factors within and outside the work place'. (*c.f.* "*A healthy working life is one that continuously provides the opportunity, ability, support and encouragement to work in ways and in an environment that allows workers to maintain and improve their health and well-being. This is not only a matter of disease or disability: it demands that every individual should be able to maximise their physical, mental and social capacity in order to gain the greatest personal benefit from their working life and to make a positive contribution to their business and society' (Scottish Executive 2004*)). The traditional risk assessment approach to health and safety is based on a biomedical model, regards work as a hazard, and aims at the primary prevention of work injury and industrial disease. This approach was developed for and remains appropriate for accidents (e.g. on building sites or with machinery), exposure to hazardous material (e.g. asbestos) or the prevention of occupational diseases (e.g. deafness or asthma). However, that approach is not appropriate and may even be counter-productive for some common health problems (e.g. 'stress', mild/moderate mental health problems, or musculoskeletal conditions) which have a high prevalence in normal people whether they are working or not, where risk factors are complex, where there is often no 'injury' or 'disease', where causation is multifactorial and ambiguous, and where primary prevention may be unrealistic. In these conditions, there is an argument that the focus should be more balanced and include improving management (both clinical and occupational) and the secondary prevention of (long-term) disability and sickness. These conditions can only be understood and managed, both clinically and at work, by addressing *all* of the personal/psychological, health-related and social/occupational factors that influence illness, disability and incapacity for work (the 'biopsychosocial model'). It is important to consider an organisational as well as an individual perspective on health at work. (*continued*)

TABLE 4: THE IMPACT OF WORK ON THE HEALTH OF PEOPLE WITH MENTAL HEALTH CONDITIONS

Authors	Key features (*Additional reviewers' comments in italics*)
Table 4c-ii: Stress: Management *continued*	
(HSE/HSL 2005) Workshop materials (Bowen *et al.* 2005) Workshop report	**Review of the risk prevention approach to occupational health: applying health models to 21st century occupational health needs** (*continued*) (UK Health & Safety Executive and Health & Safety Laboratory) Traditional risk assessment underpins occupational safety and is widely understood and accepted. It is essential to work within the current legislation and regulatory framework and not to undermine the risk assessment process. The occupational health framework must address emerging issues around health, work and well-being, but the way forward should be by evolution, enhancement and consistency. Thus, the Health and Safety strategy for the 21st century should incorporate both risk assessment and prevention *and* the broader health at work agenda. That will require a multi-disciplinary approach, which includes organisational interventions and does not only emphasise the individual. There needs to be broader collaboration between the Dept of Health, the Dept for Work and Pensions, and the HSE/HSC. Key messages are that health problems are part of everyday life and occupational health should be viewed in the wider context of health, work and well-being. It is important to emphasise the benefits and not just the hazards of work for (physical and mental) health: work may be part of the solution rather than the problem. The suitability of the person for the job ('person-job fit') is important. People of working age need to be encouraged and helped to deal with common health problems both in the occupational and non-occupational context. Messages about health and work should be framed carefully, and take a balanced approach to the relative risks and benefits.

TABLE 4: THE IMPACT OF WORK ON THE HEALTH OF PEOPLE WITH MENTAL HEALTH CONDITIONS

Authors	Key features (*Additional reviewers' comments in italics*)

Table 4c-iii: Burnout

Authors	Key features
(Lee & Ashforth 1996) Meta-analysis	**Correlates of the three dimensions of job burnout.** Meta-analysis of 61 studies of the Maslach Burnout Inventory. Emotional exhaustion correlated strongly with depersonalisation ($r=0.64$) and both correlated more weakly with personal accomplishment ($r=-0.33$ and -0.36). 5 of 8 demand variables correlated $r>0.50$ and 8 of 18 resource variables correlated $r>|0.30|$ with emotional exhaustion. 8 of 26 demand and resource variables correlated $r>|0.34|$ with depersonalisation and 3 of 26 correlated $r>|0.30|$ with personal accomplishment. (*These findings support an association between self-reports of emotional exhaustion and self-reports of psychosocial characteristics of work. They cast serious doubt on the other dimensions of the Maslach Burnout Inventory*).
(Schaufeli & Enzmann 1998) Monograph	**The burnout companion to study and practice: a critical analysis** Burnout is a metaphor: a state of exhaustion. A list is given of 132 possible affective, cognitive, physical, behavioural and motivational symptoms of burnout. Alternative definitions of burnout include: 1) a (multi-dimensional) syndrome of emotional exhaustion, depersonalisation and reduced personal accomplishment that can occur among individuals who do 'people work' of some kind; 2) a state of physical, emotional and mental exhaustion caused by long-term involvement in situations that are emotionally demanding; 3) an expectationally mediated, job-related, dysphoric and dysfunctional state in an individual without major psychopathology who has a) functioned for a time at adequate performance and affective levels in the same job situation and who b) will not recover to previous levels without outside help or environmental rearrangement. Four alternative descriptions of the 'process' of burnout are also listed. Recognises the conceptual and practical difficulties in distinguishing burnout from stress, depression and chronic fatigue, but argues these can be overcome. Diagnosis and assessment is mainly based on self-report, and 90% of studies use the Maslach Burnout Inventory. Contrary to cross-sectional studies that show a correlation between job demands and burnout, eight longitudinal studies found that work demands either had very small or non-significant effects on burnout (p95). (*Though the authors argue that is likely due to methodological problems and should not necessarily be interpreted as evidence against the causal link between high job demands and burnout*).

TABLE 4: THE IMPACT OF WORK ON THE HEALTH OF PEOPLE WITH MENTAL HEALTH CONDITIONS

Authors	Key features (Additional reviewers' comments in italics)

Table 4c-iii: Burnout continued

(Cox et al. 2005) Editorial	**The conceptualisation and measurement of burnout** 30 years of research have not yet convincingly answered many of the long-standing questions: • There are conflicting views on what burnout actually is and how it should be conceptualised. • There is lack of agreement on the nature and structure of the burnout phenomenon. • What are the psychometric properties of particular burnout instruments, and how are the different factors related? • Does burnout differ from or how does it relate to the broader and longer established concept of stress? • Is burnout different from or how does it relate to minor psychiatric illness such as anxiety or depression?

TABLE 5: THE IMPACT OF WORK ON THE HEALTH OF PEOPLE WITH MUSCULOSKELETAL CONDITIONS

Authors	Key features (*Additional reviewers' comments in italics*)

Table 5: The Impact of Work on the Health of People with Musculoskeletal Conditions

Authors	Key features
(Felson 1994) Narrative review	**Occupation-related physical factors and arthritis** Occupational physical activities over many years can induce osteoarthritis in selected joints (notably the knee and spine in miners; the hip in farmers; upper extremity joints in pneumatic drill operators). Overall, labour force participation by those with arthritis is decreased compared with those without it, both among men and women (especially when the condition is severe). People with pre-existing arthritis are especially likely to experience work disability when faced with physically demanding jobs in which they have little control over the pace or the specific physical demands of their labour. (*No data provided on the health effects of moving people with arthritis to more accommodating jobs*).
(Fordyce 1995) IASP Task Force report	**Back pain in the workplace** (*Report from a task force assembled by the International Association for the Study of Pain (IASP); it is akin to a policy paper*). It concerns the prevention of non-specific low back pain disability, and deals with worksite-based interventions to minimise disability, a program to substitute job-change flexibility for inappropriate disability assignment, early medical management, and early disability management if that fails. Recommendations: Non-specific low back pain should be re-conceptualised as a problem of activity intolerance, not a medical problem. Emphasise worksite-based interventions. Structure medical management on a time-contingent not pain-contingent basis. To reclassify as unemployed those who fail to achieve restoration of function and return to work. Establish vocational redirection programs for the unemployed. (*The basic tenet was that work is healthy and provides a level of activity conducive with prevention of long-term disability. The underlying concepts of active management (as opposed to rest) underpin modern guidelines-based approaches to the management of low back pain*) (*Also in Table 7*).

TABLE 5: THE IMPACT OF WORK ON THE HEALTH OF PEOPLE WITH MUSCULOSKELETAL CONDITIONS

Table 5: The Impact of Work on the Health of People with Musculoskeletal Conditions *continued*

Authors	Key features *(Additional reviewers' comments in italics)*
(Frank *et al.* 1996) Narrative review	**Secondary prevention of disability in occupational back pain** A review of the natural history of low back pain (LBP) and the risk factors for chronic disability, as the basis for secondary interventions to reduce the duration of occupational disability. Despite the lack of high quality RCTs, the authors conclude that there is strongly suggestive evidence for several workplace-based interventions. 1) Management retraining to more acceptance and accommodation of LBP, facilitating prompt reporting and treatment, including active rehab services at work, and the provision of modified duties. 2) Pro-active and employee-supported communication between the workplace, injured worker, health care and other involved parties. 3) 'Managed care' to ensure optimum medical treatment and rehabilitation, according to the best scientific evidence and current guidelines. 4) Integration of all these elements in a comprehensive intervention programme in the workplace.
(NIOSH 1997) Systematic review	**Musculoskeletal disorders and workplace factors** *(Large, systematic review of the epidemiological evidence on risk factors for a wide variety of work-related musculoskeletal disorders)*. Concluded that the consistently positive findings from a large number of cross-sectional studies *(which do not establish causation)*, strengthened by the limited number of prospective studies, provides strong evidence for increased risk of work-related musculoskeletal disorders for some body parts. For some body parts and risk factors there is some epidemiological evidence for a causal relationship. For other body parts and risk factors, there are insufficient studies from which to draw conclusions or the overall conclusion from the studies is equivocal. In general there is limited detailed quantitative information about exposure-response relationships between risk factors and musculoskeletal disorders. The reviewers considered that the epidemiological literature identified a number of specific physical exposures strongly associated with specific musculoskeletal disorders when exposures are intense, prolonged, and particularly when workers are exposed to several risk factors simultaneously. There is evidence that psychosocial factors related to the job and work environment play a role in the development of work-related musculoskeletal disorders of the upper extremity and back. Musculoskeletal disorders can also be caused by non-work exposures. There are insufficient studies to determine whether continued exposure to physical factors alters the prognosis of musculoskeletal disorders. *(This review does not clearly distinguish between incidence, prevalence, injury, chronicity, and work loss, and simply assumes that statistical associations represent a causal relationship. Because of the focus on risk factors as opposed to outcomes, it provides little information on work retention or return-to-work issues). (See also National Research Council 1999, and De Beek & Hermmans 2000 below).*

TABLE 5: THE IMPACT OF WORK ON THE HEALTH OF PEOPLE WITH MUSCULOSKELETAL CONDITIONS

Authors	Key features (*Additional reviewers' comments in italics*)

Table 5: The Impact of Work on the Health of People with Musculoskeletal Conditions *continued*

(Burton 1997) Narrative review	**Back injury and work loss** Took the stance that much work is physically demanding and may (frequently) lead to some discomfort and pain – these transient symptoms may be a normal consequence of life, but a proportion of people will have difficulty managing their symptoms. Control of occupational low back pain disability through ergonomic intervention, based on biomechanical principles, had so far been unhelpful. Traditional secondary prevention strategies of rest and return to restricted work duties have been suboptimal. Biomechanics/ergonomics considerations may be related to the first onset of back pain, but there was little evidence that secondary intervention based solely on these principles will influence the risk of recurrence or progression to chronicity. There was little evidence that return to 'normal' work is detrimental in terms of prolonging disability. The balance of the evidence favoured a proactive approach to rehabilitation – early return to normal work where possible along with complementary advice to reduce the risk of long-term incapacity. (*A rather idiosyncratic review, though the general principles are included in guidelines-based approaches to the management of low back pain*).
(Ferguson & Marras 1997) Narrative review	**Surveillance measures and risk factors for low back pain** Surveillance measures fall into four main types (*adapted slightly by the present reviewers*): survey of symptoms; reported injury; incidence surveillance from medical or occupational health records, lost time from work. These different surveillance measures may be viewed as a temporal or severity progression. The authors analysed a wide range of physical and psychosocial risk factors at work against these different surveillance measures, and showed that the findings depended on which surveillance measure was used. As LBP progresses from symptoms to disability, psychosocial (as opposed to physical exposure) factors play a more prominent role.
(Westgaard & Winkel 1997) Narrative review	**Guidelines for occupational musculoskeletal load as a basis for intervention** The most effective interventions were considered to be (1) 'organisational culture' using multiple interventions with high stakeholder commitment to reduce identified risk factors; and (2) modifier interventions focusing on workers at risk and using measures which actively involve the individual. However, serious methodological weaknesses mean that there is insufficient scientific evidence to draw any firm conclusions about the impact or effect sizes of these interventions. (*This review included 92 studies: they were not strictly ergonomic and very few were tested in randomised controlled trials*).

TABLE 5: THE IMPACT OF WORK ON THE HEALTH OF PEOPLE WITH MUSCULOSKELETAL CONDITIONS

Authors	Key features (Additional reviewers' comments in italics)

Table 5: The Impact of Work on the Health of People with Musculoskeletal Conditions *continued*

Authors	Key features
(Frank *et al.* 1998) Narrative review	**Preventing disability from work-related low back pain.** Synthesis of return to work approaches with focus on the stage (phase) of back pain. Management in the first 3–4 weeks should be conservative according to current clinical guidelines. Interventions at the sub-acute stage (between 3–4 and 12 weeks) should focus on return to work and can reduce time lost from work by 30–50%. There is substantial evidence that employers who promptly offer appropriately modified work can reduce the duration of work loss by at least 30%; a frequent spin-off is a reduction in the incidence of new back pain claims. The basic message was that it is important to get all the players (workers, health professionals and employers) onside. (*The concept of getting all the stakeholders onside is doubtless necessary for successful occupational management of LBP, but it is not sufficient in itself (Scheel et al. 2002c)*).
(National Research Council 1999) Workshop report	**Work-related musculoskeletal disorders** There is a strong association between biomechanical stressors at work and reported musculoskeletal pain, injury, loss of work and disability. There is a strong biological plausibility to the relationship between the incidence of musculoskeletal disorders and high-exposure occupations, but methodological weaknesses make it difficult to draw strong causal inferences or to establish the relative importance of task and other factors. Evidence that lower levels of biomechanical stress are associated with musculoskeletal disorders remains less definite. Research clearly demonstrates that reducing the amount of biomechanical stress and interventions which tailor corrective action to individual, organisational and job characteristics can reduce the reported rate of musculoskeletal disorders for workers who perform high-risk tasks.

TABLE 5: THE IMPACT OF WORK ON THE HEALTH OF PEOPLE WITH MUSCULOSKELETAL CONDITIONS

Authors	Key features (*Additional reviewers' comments in italics*)

Table 5: The Impact of Work on the Health of People with Musculoskeletal Conditions *continued*

Authors	Key features
(Buckle & Devereux 1999) Narrative review	**Work-related neck and upper limb musculoskeletal disorders** (European Agency for Safety and Health at Work) (*This narrative review was invited by the European Agency for Safety and Health at Work, with the report published following a consultation process*). There is little evidence of the use of standardised diagnostic criteria across EU member states. Understanding of the pathogenesis of these disorders varies greatly depending on the condition. There is scientific evidence for a positive relationship between the occurrence of some neck and upper limb musculoskeletal disorders and the performance of work, especially where high levels of exposure are present. Consistently reported risk factors requiring consideration in the workplace are posture (notably relating to the shoulder and wrist), force applications at the hand, hand-arm exposure to vibration, direct mechanical pressure on body tissues, effects of cold work environment, work organisation and worker perceptions of the wok organisation (psychosocial work factors) — exposure-response relationships are difficult to deduce. There is debate about the influence of repetitiveness and fatigue. It was felt that the identification of workers in the extreme exposure categories is a priority for any preventive strategy. The importance of health and risk surveillance was emphasised, and is supported by EU directives. (*With difficulties in establishing agreed pathogenesis of symptoms, the word 'disorder' may not be entirely appropriate for many of symptomatic states. The project seemingly did not include consideration of health effects of working whilst experiencing symptoms*).
(Hoogendoorn *et al.* 1999) Systematic review	**Physical load during work and leisure time as risk factors for low back pain** (*Comprehensive review of 31 longitudinal studies of the effect of physical load during work and leisure*) There is strong evidence for manual materials handling (lifting, moving, carrying and holding loads), bending and twisting, and whole-body vibration as risk factors for reporting LBP; moderate evidence for patient handling and heavy physical work; contradictory evidence for standing or walking, sitting, sports, and total leisure-time physical activity. More research is needed to determine the magnitude of the effect of the various risk factors (dose-response relationships).

TABLE 5: THE IMPACT OF WORK ON THE HEALTH OF PEOPLE WITH MUSCULOSKELETAL CONDITIONS

Table 5: The Impact of Work on the Health of People with Musculoskeletal Conditions *continued*

Authors	Key features *(Additional reviewers' comments in italics)*
(De Beek & Hermans 2000) Narrative review	**Work-related low back disorders** (European Agency for Safety and Health at Work) *(Narrative review prepared for the European Agency for Safety and Health at Work, with the report published following a number of expert workshops and consultation process).* Recognises that low back pain has a high prevalence among the population, giving evidence-linked figures of: annual incidence 4.7%; point prevalence 19%; annual prevalence ?40%. Around 25% of people with low back pain are restricted in daily activities. Only about 50% of people with low back problems seek medical advice. Notes generally favourable prognosis for most episodes, albeit symptoms are recurrent. Occupational risk factors include physical aspects of work (heavy physical work, heavy lifting, awkward postures, whole body vibration) psychosocial work-related factors (low social support and low job satisfaction) and work organisational factors (low job content, poor work organisation) – it was felt convenient to view risk in terms of a combined 'overload' on the musculo-skeletal system. Strategies to prevent low back pain include both workplace (ergonomics) and health care (rehabilitation). It was thought somewhat artificial to separate low back disorders from other musculo-skeletal disorders since a common approach is needed. *(Seemingly, the brief did not include consideration of the health effects of working whilst experiencing low back pain, yet the epidemiology clearly shows most workers with low back pain do not take time away from work).*
(Davis & Heaney 2000) Systematic review	**Psychosocial work characteristics and low back pain** *(This is the most comprehensive and methodologically critical review of psychosocial aspects of work in the context of low back pain).* There are considerable methodological weaknesses to most studies. The association between psychosocial aspects of work and low back pain is significantly weakened if physical work load is controlled for. In view of the methodological weaknesses it is difficult to draw firm conclusions. Nevertheless, there is strong evidence for a weak relationship between certain psychosocial aspects of work and reports of low back pain. Workers' self-reported, subjective reactions to psychosocial aspects of work (e.g. job dissatisfaction and job stress) are more consistently related to reported back pain than more objective aspects of work (e.g. work overload, lack of control over work, quality of relationship with co-workers).

TABLE 5: THE IMPACT OF WORK ON THE HEALTH OF PEOPLE WITH MUSCULOSKELETAL CONDITIONS

Authors	Key features (*Additional reviewers' comments in italics*)

Table 5: The Impact of Work on the Health of People with Musculoskeletal Conditions *continued*

Authors	Key features
(Abenhaim *et al.* 2000) Task force report	**Role of activity in the therapeutic management of back pain** The authors introduce a conceptual framework for the relation between back pain and occupational activity. Pain is the initiator of a series of psychological and occupational manifestations that are linked together in the biopsychosocial model. Workers with back pain may or may not experience activity limitations or restriction in employment participation. Occupational activity may be regular, reduced (activity disrupted), or interrupted (completely incapable of performing any occupational activity). In all three categories, the relation between health care, the workplace environment, and the patient is iterative (pain providing a feedback mechanism in response to medical or occupational interventions); workers with back pain have two courses of action - seek medical attention to reduce their pain, or attempt to modify their activity in the workplace, to accommodate the pain. (*Other options include complementary therapies, self-treatment, self-c ertification or simply coping*). The Task Force recommended that rest beyond the first few days of back pain (or nerve root pain) was contraindicated. Activity was considered appropriate for back pain at all stages. Work, as tolerated (perhaps with temporary modification), was considered appropriate for back pain at all stages. The importance of establishing return to regular occupational activities as soon as possible was emphasised as a therapeutic goal – this being a reflection of the necessity of minimising the duration of work absence to avoid compromising the probability of work-return.
(National Research Council 2001) Panel review	**Musculoskeletal disorders and the workplace** This US panel concluded: musculoskeletal disorders should be approached in the context of the whole person rather than focusing on body regions in isolation. There is a clear relationship between back disorders and physical load (manual handling, frequent bending and twisting, heavy physical work and whole-body vibration) and between disorders of the upper extremities and repetition, force and vibration. (*That relationship is not claimed to necessarily be causative*). Work-related psychosocial factors associated with low back disorders include rapid work pace, monotonous work, low job satisfaction, low decision latitude and job stress. Work-related psychosocial factors associated with upper extremity disorders include high job demands and high job stress. Some individual characteristics (e.g. age, psychosocial factors) affect vulnerability to work-related musculoskeletal disorders. The basic biomechanics literatures provide evidence of plausible mechanisms for the association between musculoskeletal disorders and workplace physical exposures. Modification of various physical factors and psychosocial factors could reduce the risk of symptoms for low back and upper extremity disorders. (*Essentially a 'panel consensus' document, albeit comprehensively reviewing the literature. Focused on evidence for work-relatedness of musculoskeletal disorders and the potential value of ergonomics interventions*).

TABLE 5: THE IMPACT OF WORK ON THE HEALTH OF PEOPLE WITH MUSCULOSKELETAL CONDITIONS

Table 5: The Impact of Work on the Health of People with Musculoskeletal Conditions *continued*

Authors	Key features (*Additional reviewers' comments in italics*)
(Koes *et al.* 2001) Systematic review	**International comparison of clinical guidelines for management of low back pain** There is considerable agreement across LBP guidelines from 11 countries. Consistent features were early and gradual reactivation of patients, the discouragement of bed rest, and the recognition of psychosocial factors as risk factors for chronicity. (*This review did not give details for return-to-work recommendations from the guidelines*).
(RCGP 1999) Clinical guidelines	• The UK clinical LBP guidelines from the Royal College of General Practitioners considered there was limited scientific evidence that advice to return to work within a planned short time may lead to shorter periods of work loss and less time off work. Recommendations included provision of positive messages, advice to stay at work or return as soon as possible, and consideration of reactivation/rehabilitation for those not returned to ordinary activities and work by 6 weeks.
(de Buck *et al.* 2002) Systematic review	**Rehabilitation for chronic rheumatic diseases** Work disability is a major consequence of disease in patients with chronic rheumatic diseases (rheumatoid arthritis, systemic lupus erythematosus, ankylosing spondylitis, spondylarthropathy). Work disability is substantial with rheumatic diseases (e.g. 20% to 40% of rheumatoid arthritis patients quit their job within the first 3 years of the disease), thus attention is being paid to preventing disability and promoting return to work. Five of six studies (uncontrolled) showed that 15% to 69% of patients in multidisciplinary vocational rehabilitation programs successfully return to work. Since recurrent work loss is a problem, continued access to job retention services should be considered. Because work disability and sick leave are associated with substantial inconvenience and costs for individuals and society, vocational rehabilitation programs might be considered for preventing loss of paid employment.
(Woods & Buckle 2002) Narrative review	**Work, inequality and musculoskeletal health** Reviews the following workplace and individual factors and their association with musculoskeletal ill health: social support, access to health information/education at work, job insecurity, low status work, income, education level, age, gender, and ethnicity. Gaps in knowledge, complex interrelationships and lack of independence of the variables have meant that attributing causal relationships is not possible.

TABLE 5: THE IMPACT OF WORK ON THE HEALTH OF PEOPLE WITH MUSCULOSKELETAL CONDITIONS

Authors	Key features (Additional reviewers' comments in italics)

Table 5: The Impact of Work on the Health of People with Musculoskeletal Conditions *continued*

(McClune *et al.* 2002) Narrative review	**Whiplash associated disorders** *(The review concerned a wide range of issue surrounding whiplash injuries, with the focus on deriving messages for a patient-centred educational booklet).* The key messages were: • Serious physical injury is rare in whiplash incidents • Reassurance about good prognosis is important • Over-medicalisation is detrimental • Recovery is improved by early return to normal pre-accident activities *(by implication that includes work)* • Positive attitudes and beliefs are helpful in regaining activity levels – negative attitudes and beliefs contribute to chronicity. The same general messages concerning early activation appear in the various guidance initiatives. All start with the so-called Quebec grading of severity, under which most whiplash injuries are considered minor injuries characterised by symptoms rather than objective damage. Most people return to work after whiplash (~12% remain off work after 6 months). Patients with Grade I should be able to continue work almost regardless of job requirements; Grades II and III may require some initial work loss (less so for physically non-stressful work. Non-return to work may be due to development of psychological complications. [See also (Spitzer *et al.* 1995; Allen *et al.* 1997; Motor Accidents Authority 2001; ABI 2003)]
(Staal *et al.* 2003) Systematic review	**International comparison of occupational health guidelines for management of low back pain** National occupational health guidelines from 6 countries were reviewed. All were said to be evidence-based, and there was general agreement on numerous issues fundamental to occupational health management of back pain. Psychosocial factors (both individual and workplace) can be obstacles to recovery. Advice should be given that low back pain is a self-limiting, though recurrent, condition. Prolonged work loss is detrimental. Remaining at work or an early (gradual) return to work, if necessary with modified duties, should be encouraged and supported. There is no need to wait for complete symptom resolution. *(continued)*

TABLE 5: THE IMPACT OF WORK ON THE HEALTH OF PEOPLE WITH MUSCULOSKELETAL CONDITIONS

Table 5: The Impact of Work on the Health of People with Musculoskeletal Conditions *continued*

Authors	Key features (*Additional reviewers' comments in italics*)
(Staal *et al.* 2003) Systematic review (ACC and the National Health Committee 1997) Guidance	**International comparison of occupational health guidelines for management of low back pain** (*continued*) • *This New Zealand guidance on managing back pain in the workplace is based on the idea of 'active and working'. The focus is on 'management' because LBP, being so common and often not caused by work, is almost impossible to prevent.* • The best treatment is to stay active and at work – with temporary modifications if needed. • There are many factors, physical and non-physical, that can affect return to work • The workplace has a key role to play in helping people stay at work or return early. • Employer support and encouragement to work can speed recovery – just waiting until symptom free or leaving everything to the health professional can slow it down. • Communication between the worker, the employer and the health professional is pivotal to the active-and-working approach.
(Carter & Birrell 2000), (Waddell & Burton 2001) Guidelines & evidence review	The UK Occupational health guidelines included the following (*key*) recommendations: • LBP is common and frequently recurrent - physical demands of work are only one factor influencing LBP – prevention and case management need to be directed at both physical and psychosocial factors. • LBP is not a reason for denying employment in most circumstances – care should be taken when placing individuals with a strong history in physically demanding jobs. • Advise on good working practices such as specified in manual handling regulations. • Encourage workers with LBP to continue as normally as possible and to remain at work, or to return to work at an early stage, even if they still have some LBP – consider temporary adaptation of the job or pattern of work if necessary.

TABLE 5: THE IMPACT OF WORK ON THE HEALTH OF PEOPLE WITH MUSCULOSKELETAL CONDITIONS

Authors	Key features (*Additional reviewers' comments in italics*)

Table 5: The Impact of Work on the Health of People with Musculoskeletal Conditions *continued*

Authors	Key features
(Schonstein *et al.* 2003) Systematic review	**Physical conditioning/work hardening programmes for workers with back and neck pain (Cochrane review)** Work-oriented pain management programs aim to help people return to work and improve work abilities. The programs (variously called work or physical conditioning, work hardening or functional restoration) sometimes simulate work tasks and include physical and muscle training exercises that improve physical condition and well-being. The reviewers' conclusions were: Physical conditioning programs that include a cognitive-behavioural approach plus intensive physical training (specific to the job or not) that includes aerobic capacity, muscle strength and endurance, and coordination, are in some way work-related, and are given and supervised by a physiotherapist or a multidisciplinary team, seem to be effective in reducing the number of sick days for some workers with chronic back pain, when compared with usual care. There was little evidence that specific exercise programs that did not include a cognitive-behavioural component had any effect on time lost from work.
(Proper *et al.* 2003) Systematic review	**Worksite physical activity programs and physical activity, fitness and health** Fifteen randomised trials and 11 non-randomised trials of high quality. Strong evidence was found for positive effect of a worksite physical activity program on physical activity and musculoskeletal disorders. Limited evidence was found for a positive effect on fatigue. For physical fitness, general health, blood serum lipids, and blood pressure, inconclusive evidence or no evidence was found for a positive effect. To increase the level of physical activity and to reduce the risk of musculoskeletal disorders, the implementation of worksite physical activity programs is supported.

TABLE 5: THE IMPACT OF WORK ON THE HEALTH OF PEOPLE WITH MUSCULOSKELETAL CONDITIONS

Authors	Key features (Additional reviewers' comments in italics)

Table 5: The Impact of Work on the Health of People with Musculoskeletal Conditions *continued*

Authors	Key features
(Waddell & Burton 2004) Review of reviews	Concepts of rehabilitation for common health problems (*The review covered the range of common health problems but only information related to musculoskeletal disorders is noted here*) • Low back pain: Advice to stay active and continue ordinary activities (including work) as normally as possible despite pain leads to faster return to work, fewer recurrences and less work loss over the following year than more passive approaches. Most workers with back pain are able to continue working or to return to work within a few days or weeks, even if they still have some residual or recurrent symptoms. • Other musculoskeletal disorders: There seems to be common strands to the different musculoskeletal symptoms/disorders. The themes were broadly consistent with back pain (where there is a much higher quantity of evidence) and there was nothing contrary to the evidence on back pain. • Modified work: Helpful for assisting return to work for back pain and other musculoskeletal disorders. Modified work should be a temporary measure to accommodate reduced capacity; it facilitates early return to normal duties, assuming the risks are suitably assessed and controlled – assignment to permanent modified work can be harmful. (*Also in Table 3*).
(National Health and Medical Research Council 2004) Clinical guidelines	**Management of acute musculoskeletal pain** Australian evidence-based clinical guidelines for management of a variety of painful musculoskeletal conditions. Conditions covered comprise: acute low back pain, acute thoracic pain, acute neck pain, acute shoulder pain, acute knee pain. (*Occupational issues and return to work were not the focus of this guidance, but the recommendations regarding activity are of relevance to work*). For low back pain, advice to stay active reduces sick leave compared to bed rest (as well as having small benefits for pain and function). For thoracic pain, it is, in general, important to resume normal activities as soon as possible. For neck pain, encouraging the resumption of normal activities and movement of the neck is more effective than a collar and rest. For shoulder pain, although pain may make it difficult to carry out usual activities, it is important to resume normal activities as soon as possible. For knee pain, maintenance of normal activity has beneficial effect on patellofemoral pain compared to no treatment or use of orthoses.

TABLE 5: THE IMPACT OF WORK ON THE HEALTH OF PEOPLE WITH MUSCULOSKELETAL CONDITIONS

Table 5: The Impact of Work on the Health of People with Musculoskeletal Conditions *continued*

Authors	Key features (*Additional reviewers' comments in italics*)
(COST B13 working group 2004) European clinical and prevention guidelines	**European guidelines for management of low back pain** The guidelines were based on systematic evidence reviews in three areas: management of acute low back pain, management of chronic low back pain, and prevention in low back pain. • The clinical guidelines for acute LBP considered there was evidence that advice to stay active led to less sick leave and less disability. There was consensus that advice to stay at work or return to work if possible is important. Longer duration of work absenteeism is associated with poor recovery (lower chance of ever returning to work). An appendix on back pain at work included information and recommendations taken from various occupational health guidelines (see Staal et al 2003 above). • The clinical guidelines for chronic LBP noted that after an initial episode of LBP, 44-78% people have relapses of pain and 26-37% experience relapses of work absence. In workers having difficulty returning to normal occupational duties at 4-12 weeks, the longer a worker is off work with LBP the lower the chances of ever returning to work. Intensive physical training ("work hardening") programs with a cognitive-behavioural component are more effective than usual care in reducing work absenteeism. • The guidelines for prevention in low back pain suggest that the general nature and course of commonly experienced LBP means that there is limited scope for preventing its incidence (first-time onset). Primary causative mechanisms remain largely undetermined: risk factor modification will not necessarily achieve prevention. Nevertheless, there is evidence suggesting that prevention of various consequences of LBP (e.g. recurrence, care seeking, disability, and workloss) is feasible. Overall, there is limited robust evidence for numerous aspects of prevention in LBP; for interventions where there is acceptable evidence, the effect sizes are rather modest. For workers with or without back pain the following statements are made: (1) physical exercise is recommended in the prevention of LBP, for prevention of recurrence of LBP, and for prevention of recurrence of sick leave due to LBP; (2) temporary modified work and ergonomic workplace adaptations can be recommended to facilitate earlier return to work for workers sick listed due to LBP; (3) there is insufficient consistent evidence to recommended physical ergonomics interventions alone for prevention in LBP; (4) there is insufficient consistent evidence to recommend stand-alone work organisational interventions; (5) multidimensional interventions at the workplace can be recommended in principle.

TABLE 5: THE IMPACT OF WORK ON THE HEALTH OF PEOPLE WITH MUSCULOSKELETAL CONDITIONS

Authors	Key features (*Additional reviewers' comments in italics*)

Table 5: The Impact of Work on the Health of People with Musculoskeletal Conditions *continued*

Authors	Key features
(Helliwell & Taylor 2004) Narrative review	**Repetitive strain injury** Pain in the forearm is common in the community. In the workplace it is associated with frequent high repetition, high forces, prolonged abnormal postures, and psychosocial issues. Early intervention and active management is important: the principles of the well-developed back pain guidelines apply – reassurance (addressing psychosocial factors), maintain work if possible, temporary activity modification. Ergonomic interventions may make the workplace more comfortable, and may reduce sickness absence. (*Focus was mostly on clinical issues*).
(Punnett & Wegman 2004) Narrative review	**Work-related musculoskeletal disorders – epidemiology and the debate** The debate about the work-relatedness of musculoskeletal disorders reflects both confusion about epidemiological principles and gaps in the scientific literature. Some dispute remains over the relative importance of physical ergonomic risk factors. This paper is said to address the controversy with reference to the report from the National Research Council (2001). The authors consider the available epidemiological evidence to be substantial, but accept more research is needed concerning the latency effect, natural history, prognosis, and potential for selection bias in the form of the healthy worker effect. Examination techniques still do not exist that can serve as a gold standard for many of the symptoms commonly reported in workplace studies. Exposure assessment has too often been limited to crude indicators such as job title, and lack of standardised exposure measures limits ability to compare studies. Despite these challenges, the epidemiological literature on work-related musculoskeletal disorders in combination with extensive laboratory evidence of pathomechanisms related to work stressors is convincing to most (*sic*). (*The authors' underlying tenet seems to be that the case for the work-related aetiology of musculoskeletal disorders would be strengthened by research involving improved methodology and metrics. However, (logically) it seems equally possible that the reverse may be found. As important as the underlying data is the way it is interpreted – that part of the debate also remains unresolved*).

TABLE 5: THE IMPACT OF WORK ON THE HEALTH OF PEOPLE WITH MUSCULOSKELETAL CONDITIONS

Authors	Key features (*Additional reviewers' comments in italics*)

Table 5: The Impact of Work on the Health of People with Musculoskeletal Conditions *continued*

(Backman 2004) Narrative review	**Work disability in rheumatoid arthritis** Approximately one-third of people with rheumatoid arthritis will leave employment prematurely. *Work disability results from a complex interaction of characteristics of individuals, the nature of their work, and their environment, including the physical workplace, policies related to work accommodation, and personal relationships. Early assessment of possible work limitations and potential for vocational rehabilitation should be considered in the evaluation of employed patients and those wishing to work.*
(de Croon *et al.* 2004) Systematic review	**Prediction of work disability in rheumatoid arthritis** Work disability is a common outcome in rheumatoid arthritis (RA), and is a societal and individual problem (financial costs, loss of status). Strong evidence showed that physical job demands, low functional capacity, old age, and low education predict work disability in RA. Remarkably, biomedical variables did not consistently predict work disability. It was concluded that work disability in RA is a biopsychosocially determined misfit between individual capability and work demands. Although work disability increases during the course of the disease, there was no consistent evidence that disease duration predicts disability. There was evidence that work disability itself may stimulate disease progression, because of the loss of psychosocial, financial and medical benefits. Drug treatments were not included in the review, but it was considered that treatment with disease modifying agents may influence work disability substantially.

TABLE 5: THE IMPACT OF WORK ON THE HEALTH OF PEOPLE WITH MUSCULOSKELETAL CONDITIONS

Table 5: The Impact of Work on the Health of People with Musculoskeletal Conditions *continued*

Authors	Key features (*Additional reviewers' comments in italics*)
(ARMA 2004) Standards	**Standards of care** (UK Arthritis and Musculoskeletal Alliance) *(Derived from working groups and consultation. The Standards are intended to inform health care policy makers, and cover back pain, osteoarthritis and inflammatory arthritis. Although focused on care services, the Standards do include work issues).* The high economic impact of back pain, osteoarthritis and inflammatory arthritis is acknowledged in respect of sickness absence and disability as well as health care. The following Standards in respect of work are set down: • <u>Back pain</u>: People with back pain should be encouraged and supported to remain in work or education wherever possible – vocational rehabilitation should be available to support people in staying in existing employment or finding new employment. • <u>Osteoarthritis</u>: People with joint pain or osteoarthritis should be encouraged to remain in work or education wherever possible. Vocational rehabilitation should be available to support people staying in existing employment or finding new employment. • <u>Inflammatory arthritis</u>: People should be supported to remain in or return to employment and/or education, through access to information and services such as occupational therapy, occupational support and rehabilitation services.
(Woods 2005) Narrative review	**Work-related musculoskeletal health and social support** Concerns the relationship between the level of social support at work (e.g. poor communication channels, unsatisfactory work relationships, unsupportive organisational culture) and work-related musculoskeletal ill-health (reported symptoms, sick leave, medical consultation, disability retirement). Indicates a lack of social support (from co-workers, supervisors or managers) is a risk factor for musculoskeletal ill-health *(though not necessarily causative)*. In addition, there is limited evidence that poor social support is associated with musculoskeletal absence, restricted activity, and not returning to work after a musculoskeletal problem. Prevention programmes should involve psychosocial as well as ergonomic elements. The question of whether social support causes musculoskeletal disorders or affects behaviour of patients with existing musculoskeletal disorders requires further clarification. (*The findings are based on cross-sectional, case-control studies and prospective research*).

TABLE 5: THE IMPACT OF WORK ON THE HEALTH OF PEOPLE WITH MUSCULOSKELETAL CONDITIONS

Authors	Key features (*Additional reviewers' comments in italics*)

Table 5: The Impact of Work on the Health of People with Musculoskeletal Conditions *continued*

Authors	Key features
(Walker-Bone & Cooper 2005) Narrative review	**Occupational associations of soft tissue musculoskeletal disorders of neck and upper limb** Concern was occupational associations with neck and upper limb musculoskeletal disorders. Considered separately neck disorders, shoulder disorders, epicondylitis, non-specific forearm pain, and carpal tunnel syndrome. • Neck disorders: High background prevalence of neck pain among adults in developed countries (point prevalence up to 34%); contributes to sickness absence and demands on medical services. Neck pain and neck disorders are associated with mechanical and psychosocial workplace factors (with complex interactions) – preventive strategies are not convincing. • Shoulder disorders: High background prevalence of shoulder pain (point prevalence up to 26%). Symptoms/disorders are associated with overhead work and possibly repetitive work: occupational psychosocial factors are also implicated (this holds true even when the outcome studied is a specific diagnosis). • Epicondylitis: Strenuous manual tasks seem to be associated with epicondylitis, but unclear if mechanical factors initiate the disorder or aggravate a tendency among predisposed people: emerging evidence suggesting association with psychosocial factors. • Non-specific forearm pain: Rare among working age adults (point prevalence 0.5%). Significantly associated with psychological distress but not with any mechanical exposures. • Carpal tunnel syndrome: Aetiology controversial due to problem of case definition. Overall, workplace factors may be contributory (force, repetition and vibration). Neck and upper limb pain is a common problem among working age adults and contributes to sick leave. Workplace factors such as prolonged abnormal posture and repetition contribute to these conditions. Psychosocial influences show the aetiology is complex, and both types of factor may be important, though there is insufficient evidence to determine the relative contribution. (*The odds ratios quoted from the original studies tended to be <2 for physical factors and >3 for psychosocial factors.*)

TABLE 5: THE IMPACT OF WORK ON THE HEALTH OF PEOPLE WITH MUSCULOSKELETAL CONDITIONS

Authors	Key features (*Additional reviewers' comments in italics*)
Table 5: The Impact of Work on the Health of People with Musculoskeletal Conditions *continued*	
(Henriksson *et al.* 2005) Narrative review	**Women with fibromyalgia** *(A review of literature focused on the work status of women with fibromyalgia).* Fibromyalgia is predominantly reported by, and diagnosed in, females. Prevalence 1–5%. Limitations caused by pain, fatigue, decreased muscle strength and endurance influence work capacity. Studies from various countries show many women with fibromyalgia (34% to 77%) remain at work even after many years with pain. When individual adjustments in the work situation (reduced, more flexible hours) are made, they can continue to work and find satisfaction in their work role. It seems that when the women find a work situation that matches their ability, they continue to work. The total life situation, other commitments, type of work tasks, ability to influence work situation, and the physical and psychosocial work environment are important in determining whether a person can remain in work.
(Loisel *et al.* 2005) Narrative review and expert opinion	**Prevention of work disability due to musculoskeletal disorders** The evidence shows that some clinical interventions (advice to return to modified work and graded activity programmes) and some non-clinical interventions (at a service and policy/community level but not at practice level) are effective in reducing work absenteeism. Implementation of evidence on work disability is problematic. The limited implementation of these evidence-based practices may be related to the complexity of the problem, as it is subject to multiple legal, administrative, social, political and cultural challenges.
(Cairns & Hotopf 2005) Systematic review	**Prognosis of chronic fatigue syndrome** Aim of the review was to identify occupational outcomes in chronic fatigue syndrome. The median full recovery rate during variable length follow-up was 5%, but the symptom improvement rate was 40%. Less fatigue severity at baseline, a sense of control over symptoms and not attributing illness to a physical cause were associated with a good outcome. Return to work at follow-up varied from 8% to 30%. It is indisputable that it is easier to return to work after shorter periods of sickness absence – services should be available to provide early treatment and rehabilitation. Medical retirement should be postponed until a trial of such treatment has been given.

TABLE 5: THE IMPACT OF WORK ON THE HEALTH OF PEOPLE WITH MUSCULOSKELETAL CONDITIONS

Authors	Key features (*Additional reviewers' comments in italics*)
Table 5: The Impact of Work on the Health of People with Musculoskeletal Conditions *continued*	
(Franche *et al.* 2005) Systematic review	**Workplace-based return-to-work interventions.** There was strong evidence that work disability duration is significantly reduced by work accommodation offers and contact between healthcare provider and workplace; and moderate evidence that it is reduced by interventions which include early contact with worker by workplace, ergonomic work site visits, and presence of a return-to-work coordinator. For these five intervention components, there was moderate evidence that they reduce costs associated with work disability duration. There was limited evidence on the sustainability of these effects. There was mixed evidence regarding direct impact on quality-of-life outcomes. (*Importantly, however, this review found no evidence that return to work had adverse impact on quality of life*). (*Also in Table 3*).
(Staal *et al.* 2005) Narrative review	**Physical exercise to improve disability and return to work in low back pain** Acknowledges the body of literature indicating that physical exercise might be effective to stimulate return to work and improve function in workers who are absent from work due to low back pain. However, in cases of occupational low back pain, it is often a physical incident or activity that is blamed for the precipitation of back pain or sciatica and held responsible for damaging spinal structures. This review explores the literature to determine whether the risk of additional back pain and work absence increases in people with a history of back pain, if they resume physical activities including exercise and work. Physical exercises are not associated with an increased risk for recurrences. Authors consider staying active and increasing the level of physical activity are safe, despite increased loading of spine structures.

TABLE 5: THE IMPACT OF WORK ON THE HEALTH OF PEOPLE WITH MUSCULOSKELETAL CONDITIONS

Authors	Key features (Additional reviewers' comments in italics)

Table 5: The Impact of Work on the Health of People with Musculoskeletal Conditions *continued*

Authors	Key features
(Talmage & Melhorn 2005)	**Working with common musculoskeletal problems** (American Medical Association) *(American Medical Association guide book for primary care physicians and care providers to assist navigation of return to work issues, supported by science and consensus: the authors admit a firm belief that work is good for man).*
(Haralson 2005) Physician guidance	• <u>Common lower extremity problems</u>: Lower extremity injuries are a common cause of the loss of the ability to work. With proper job accommodations (e.g. redesign to reduce the need for locomotion) people with lower extremity problems can return to work relatively quickly (with general health benefit) and there seems little reason to keep workers with most lower extremity problems off work for extended periods. *(A chapter from a guide book primarily to present evidence-based advice to physicians involved in return to work assessments – the approach involves consideration of risk, capacity and tolerance at the individual level).*
(Melhorn 2005) Physician guidance	• <u>Common upper extremity problems</u>: Returning an individual with an upper extremity problem to work requires a balance between the demands of the job and the capability of the patient. Temporary workplace advice on accommodations and tolerance should focus on an early return to work and improve the outcome for work-related injuries, and advance the patients' quality of life. *(A chapter from a guide book primarily to present evidence-based advice to physicians involved in return to work assessments – the approach involves consideration of risk, capacity and tolerance at the individual level).*
(Sherrer 2005) Physical guidance	• <u>Common rheumatological disorders</u>: Rheumatological disorders are varied, but they uniformly have a negative impact on work. Emerging data suggest that the majority of patients can continue to work with certain parameters, and will need aggressive control of disease activity and pain, along with appropriate workplace adaptations. *(A chapter from a guide book primarily to present evidence-based advice to physicians involved in return to work assessments – the approach involves consideration of risk, capacity and tolerance at the individual level). (Entries in respect of working with cardiorespiratory conditions are in Table 6).*

TABLE 5: THE IMPACT OF WORK ON THE HEALTH OF PEOPLE WITH MUSCULOSKELETAL CONDITIONS

Authors	Key features (Additional reviewers' comments in italics)

Table 5: The Impact of Work on the Health of People with Musculoskeletal Conditions *continued*

Authors	Key features
(D'Souza *et al.* 2005) Systematic review	**Occupational factors and lower extremity musculoskeletal and vascular disorders** The epidemiological literature on lower extremity musculoskeletal disorders, vascular disorders and occupational mechanical factors is relatively sparse compared to low back and upper extremity literature. Most of the literature concerned osteoarthritis of the hip and knee, but it was focused on surgical (i.e. severe) cases, which limits generalisability. Most of the studies on other conditions were cross-sectional and used questionable exposure assessment: more and better research is needed to examine any causal pathways.
(Burton *et al.* 2006) Systematic review	**Employment and rheumatoid arthritis** The 38 studies included in the review concerned subjects ≥18 with a diagnosis of rheumatoid arthritis (RA) and had a measure of work productivity loss (work loss and work disability). Rates of work disability in RA were similar in the USA and Europe, despite differences in social systems. Times from disease onset until 50% probability of being permanently work disabled varied from 4.5 to 22 years. Work loss was experienced by a median 66% (range – 36-84), for a median duration of 39 days (range 7-84). Baseline characteristics consistently predictive of subsequent work disability were a physically demanding work type, more severe RA and older age. An apparent decrease in the prevalence of RA-related work disability since the 1970s may be due to a decrease in physically demanding work. (*Over this time, there have also been generally increased employment rates among people with disabilities. Overall, this study shows that people with RA can work for considerable periods after disease onset and that some experience only modest work loss, with a suggestion that participation will depend, at least in part, on the physical nature of the work*).

TABLE 5: THE IMPACT OF WORK ON THE HEALTH OF PEOPLE WITH MUSCULOSKELETAL CONDITIONS

Authors	Key features (*Additional reviewers' comments in italics*)

Table 5: The Impact of Work on the Health of People with Musculoskeletal Conditions *continued*

Authors	Key features
(IIAC 2006) UK Legislation	**Prescribed diseases** (Industrial Injuries Advisory Council) The UK law provides for payment of benefits to people who are suffering from certain diseases contracted in the course of certain types of employment. These diseases are referred to as prescribed diseases (PDs) and are listed in Regulations. There is no entitlement to benefit in respect of a disease if it is not listed in the Regulations, or if the person's job is not listed against the particular disease. This is especially important for diseases common in the population at large, where it is known that some workers would have got the disease whatever job they did. A disease can only be prescribed if the risk to workers in a certain occupation is substantially greater than the risk to the general population, and the link between the disease and the occupation can be established in each individual case or presumed with reasonable certainty. In diseases which occur in the general population (e.g. chronic bronchitis and emphysema) there may be no difference in the pathology or clinical features to distinguish an occupational from a non-occupational cause. In these circumstances, in order to recommend prescription, IIAC looks for consistent evidence that the risk of developing the disease is more than doubled in a given occupation. There are a number of common musculoskeletal disorders that are considered prescribed diseases: cramp of the hand or forearm due to repetitive movements; subcutaneous cellulitis of the hand due to manual labour causing severe friction or pressure; bursitis or subcutaneous cellulitis at the knee due to severe prolonged external friction or pressure; bursitis or cellulitis at the elbow due to severe or prolonged external friction or pressure; traumatic inflammation of the tendons (tenosynovitis) affecting the hand due to manual labour or frequent or repeated movements of the hand or wrist; vibration white finger and carpal tunnel syndrome related to use of hand-held vibrating tools; osteoarthritis of the hip in agriculture as a farmer or farm worker for a period 10 years. (*Whilst it is recognised that certain exposures in certain jobs are related to certain musculoskeletal diseases, it is not implied as inevitable that exposure to the job will result in the disease*). (*Also in Table 6*).

TABLE 6: THE IMPACT OF WORK ON THE HEALTH OF PEOPLE WITH CARDIO-RESPIRATORY CONDITIONS

Authors	Key features (*Additional reviewers' comments in italics*)

Table 6a-i: Cardiac conditions - impact of work

(Schnall *et al.* 1994) Quasi-systematic review	**Job strain & cardiovascular disease** Reviews 36 studies on the relationship between job strain and cardiovascular disease outcomes (e.g. myocardial infarction, mortality) and cardiovascular disease risk factors (e.g. hypertension). Concludes that a body of literature had accumulated (*at that time*) to strongly suggest a causal association between job strain and cardiovascular disease. Several biological mechanisms for the association, notably elevated blood pressure, have received empirical support. However, it was considered the job demands-control (job strain) model has conceptual limitations that need to be addressed in further research, including other aspects of job demands (e.g. cognitive demands, workplace social support, latitude, promotion opportunities). In summary, the authors concluded that job strain is associated with a range of adverse health outcomes, including psychological strain, such as exhaustion or depression, hypertension, and various forms of cardiovascular disease. (*The reviewed studies included a mix of cross-sectional, case control and cohort studies: the quoted effect sizes were variable and largely modest*).
(Kaplan & Keil 1993) Narrative review	**Socioeconomic factors and cardiovascular disease** (*General social background*) Socioeconomic status, sometimes referred to as 'social class', covers a wide range of measures including education, income, occupation, living conditions, income inequality, and many other socio-economic aspects of life. This review found a substantial body of evidence of a consistent association between (lower) socio-economic status and the aetiology and progression of cardiovascular disease, which may be mediated by standard biologic cardiovascular risk factors and/or by psychosocial factors. Associations between all-cause mortality and education level, income level, occupational group, composite indices of these measures, poverty, unemployment, living conditions and standard of living have been demonstrated using both individual and aggregate data. A large number of studies have demonstrated associations between socio-economic status and cardiovascular risk factors (physical activity, smoking, obesity, haemostatic factors and hypertension), coronary heart disease, cardiovascular mortality (particularly coronary mortality) and mortality trends. Analysis of the association between employment status and health must distinguish those who are able to work but unable to find employment and those who are unable to work for health reasons. (*Provides little evidence or discussion on the impact of work or worklessness on cardiovascular disease*).

TABLE 6: THE IMPACT OF WORK ON THE HEALTH OF PEOPLE WITH CARDIO-RESPIRATORY CONDITIONS

Authors	Key features (Additional reviewers' comments in italics)

Table 6a-i: Cardiac conditions – impact of work continued

Authors	Key features
(Pickering 1997) Narrative review	**Occupational stress and blood pressure in men and women** Human hypertension is the end result of a number of genetic and environmental influences, and typically develops gradually over many years. The extent to which increased sympathetic activity (which plays a role in the early stage) may be the result of environmental stress is uncertain. Human epidemiological studies have shown that the prevalence of hypertension is strongly dependent on social and cultural factors. Blood pressure tends to be highest at work, and studies using ambulatory monitoring have shown that occupational stress, measured as job strain, can raise blood pressure in men, but not women (possibly associated with mens' increased left ventricular mass). The diurnal blood pressure pattern in men with high strain jobs shows a persistent elevation throughout the day and night, which is consistent with the hypothesis that job strain is a causal factor in the development of human hypertension.
(van der Doef & Maes 1998) Systematic review	**The Job Demand-Control-(Support) Model and physical health outcomes** Review of 51 studies, mainly of cardiovascular disease (CVD), in which the 'strain' model (demand-control) predominates. In contrast, the 'buffer' model (in which control buffers the impact of demands) is most prevalent in studies of self-reported psychosomatic complaints. One out of two studies of all cause mortality supported the strain model. Three out of 7 studies on CVD mortality, 7 out of 12 studies on CVD morbidity, and 3 out of 7 studies on CVD symptomatology supported the strain hypothesis (*though no data were presented on the strength of these effects.*) The review authors concluded that 'working in a high strain job appears to be associated with an elevated risk for cardiovascular disease (*though these results might be more accurately summarised as 'there is conflicting evidence that job strain is associated with any increased risk for cardiovascular disease, but no evidence from this review on the extent of such a risk and no clear evidence on cause and effect'*).

TABLE 6: THE IMPACT OF WORK ON THE HEALTH OF PEOPLE WITH CARDIO-RESPIRATORY CONDITIONS

Authors	Key features (*Additional reviewers' comments in italics*)

Table 6a-i: Cardiac conditions - impact of work *continued*

Authors	Key features
(Tsutsumi & Kawakami 2004) Systematic review	**The Effort-Reward Imbalance model** Review of 45 studies. Two case-control and seven cohort studies on cardiovascular disease showed associations between effort-reward imbalance or over-commitment and acute myocardial infarction or cardiac mortality (*which were generally significant at P <0.05, but without Bonferroni adjustment for multiple comparisons*) with odds ratios generally from 1.2 to 4.5 (*though many of those in the higher range were barely significant because of small sample sizes and there was no meta-analysis to combine these results*). Eight cross-sectional studies and two cohort studies showed similar associations between effort-reward imbalance or over-commitment and hypertension or other biological risk factors (*with all the same qualifications, as above*). (*Also in Table 4*).
(van Vegchel *et al.* 2005) Systematic review	**The Effort-Reward Imbalance model** Review of 45 studies, 24 of which were of cardiovascular disease (CVD). Thirteen out of 17 studies of cardiac symptoms/risk factors were positive with odds ratios ranging from 1.23 to 6.71, and all 8 studies of CVD morbidity or mortality were positive with odds ratios ranging from 1.22 to 8.98 (*no meta-analysis to combine these results*). (*These studies were almost entirely in men*). (*Also in Table 4*).
(Hemingway & Marmot 1999) Systematic Review	**Psychosocial factors in the aetiology of coronary heart disease** In healthy populations, prospective cohort studies show a possible aetiological role for type A/hostility, depression/anxiety, psychosocial work characteristics, and social support. In populations of patients with coronary heart disease, prospective studies show a prognostic role only for depression/anxiety, psychosocial work characteristics, and social support. Whilst psychosocial factors (particularly depression and social support) are independent aetiological and prognostic factors for coronary heart disease, there is conflicting evidence on whether psychosocial interventions reduce mortality after myocardial infraction. (*In summary, there is conflicting evidence that psychosocial aspects of work such as high demands and low control are weak risk factors for coronary heart disease; there is strong evidence that anxiety, depression and social support are more powerful risk factors*).

TABLE 6: THE IMPACT OF WORK ON THE HEALTH OF PEOPLE WITH CARDIO-RESPIRATORY CONDITIONS

Authors	Key features (*Additional reviewers' comments in italics*)
Table 6a-ii: Cardiac conditions - management	
(Shanfield 1990) Narrative review	**Return to work after myocardial infarction** The rate of return to work after an acute myocardial infarction is decreased among previously working women, blue-collar workers particularly with physically strenuous jobs, and persons with emotional problems. Although more severe infarctions decrease return to work rates, psychosocial factors appear to be more prominent in their effects on the rates. Supportive psychotherapy as well as specific advice to return to work for patients with uncomplicated acute myocardial infarction shortens length of convalescence. Little evidence exists, however, that current interventions largely geared to improving cardiac status have any impact on ultimate return to work. Specific interventions tailored to individuals at risk of not working may increase rates of return to work.
(Horgan *et al.* 1992) Report	**Working party report on cardiac rehabilitation in UK** Cardiac rehabilitation should restore patients to their optimal physiological, vocational, and social status. Return to work (RTW) is usually an explicit aim of cardiac rehabilitation. Vocational services need to be involved early in the rehab process because prolonged inactivity can lead to permanent incapacity. To this end, job characteristics should be evaluated. Good communications between the medical staff and those providing vocational counselling about medical status, exercise tolerance, and psychological outcome are essential in the decision to return to work. Resumption of work is largely determined by factors that cardiac rehab cannot influence – age, severity of disease, educational level, and adequacy of pension and retirement benefits. The role of vocational rehabilitation in recovery needs greater recognition.
(Brezinka & Kittel 1995) Narrative review	**Psychosocial factors of coronary heart disease in women** Much less research is carried out in this area in women than in men. For women, low social class, low educational attainment, the double loads of work and family, chronic troubling emotions and lack of social support emerge as risk factors. Women working before a coronary incident are more likely to withdraw from the labour market than men, and those who return to work do so later than men. Generally, return to work after a coronary incident decreases emotional distress and enhances well-being, whereas forced retirement leads to negative changes in emotional and social adjustment. Thus physicians should avoid advising female patients not to return to work – the psychosocial consequences of withdrawal might be worse than those of return to work.

TABLE 6: THE IMPACT OF WORK ON THE HEALTH OF PEOPLE WITH CARDIO-RESPIRATORY CONDITIONS

Authors	Key features (*Additional reviewers' comments in italics*)

Table 6a-ii: Cardiac conditions - management *continued*

(Lusk 1995) Narrative review	**Return to work after myocardial infarction** Most individuals want to return to work (RTW) following myocardial infarction. Overprotection at the workplace is undesirable. Occupational health nurses have an important role in planning for and implementing RTW. Specific suggestions for facilitating return included involvement early in the worker's recuperative period, including early contact with the worker, offering support and encouragement by telephoning/visiting, maintaining the link to the workplace, encourage co-workers, supervisors, and/or subordinates to maintain contact with the worker. It was considered these suggestions debunk the myth that overprotection results in 'cardiac invalids', and that return to work was partly predicted by perceptions of early high social support. In addition, occupational health nurses have a role in devising RTW programs, including modified work, which results in a reduction of dependence on physicians for determination of RTW, who may not have as clear an understanding of the physiological aspects of the job. (*A review of two just original studies, from the perspective of occupational health nursing*).
(Wenger *et al.* 1995) Guideline	**Clinical practice guideline for cardiac rehabilitation services - USA** Symptomatic and functional improvement in survivors of myocardial infarction and revascularisation procedures correlate poorly with RTW and general resumption of pre-illness lifestyle; psychosocial status appears to be a more important determinant. Exercise training exerts less influence on rates of RTW than many non-exercise variables including employer attitudes, prior employment status, and economic incentives. Exercise training, as a sole intervention, is insufficient to facilitate return to work. Education, counselling, and behavioural interventions have not been shown to improve rates of RTW, which are contingent on many social and policy issues.

TABLE 6: THE IMPACT OF WORK ON THE HEALTH OF PEOPLE WITH CARDIO-RESPIRATORY CONDITIONS

Authors	Key features (Additional reviewers' comments in italics)

Table 6a-ii: Cardiac conditions - management continued

Authors	Key features
(Dafoe & Cupper 1995) Narrative review	**Vocational considerations and return to work** Vocational issues for cardiac patients increasingly carry an expectation that cardiac patients should, if at all possible, be gainfully employed. A successful return to work can be viewed as a major milestone in rehabilitation process. Clinically, some individuals should not return, and people in their late 50s/early 60s may choose not to return if they can retire without undue financial hardship. When work return is an accepted goal, early proactive approach important. Health professionals need to anticipate, identify, and modify obstacles to return to work at an early stage, and alleviate associated anxiety. Continuing contact between employer and absent employee may facilitate RTW. Involvement of company physician in modified work is important. Co-workers' attitudes to the returning employee should be supportive, as indeed should the attitudes of the family — they need appropriate information. The (early) decision to return to work is a crucial step, needing good communication between patient, employer, primary care physician and rehabilitation service. Vocational counselling, as a component of cardiac rehabilitation, should be flexible and community-based. (*Not strictly evidence-based, rather a discussion of concepts*).
(Thompson et al. 1996) Narrative review	**Cardiac rehabilitation in UK** Patients should be encouraged to remain independent, and should have a say in what they are willing to do. RTW is considered a major end point in cardiac rehabilitation. A high percentage (62-92%) of patients who were working before myocardial infarction will return to work, but a high proportion will leave work again or change jobs in the first year after returning. Except in the case of patients with persisting angina or heart failure, failure to return to work is more often due to psychological or financial considerations than to physical constraints. Even in the case of occupations where health standards are prescribed by statute (e.g. driving, aviation), there is an increasing move towards individual assessment of risk and capacity. There needs to be collaboration between cardiac rehabilitation and occupational health medicine to ensure optimum and effective RTW. (*Summary of a workshop convened to prepare clinical guidelines and audit standards in cardiac rehabilitation in the UK*).

TABLE 6: THE IMPACT OF WORK ON THE HEALTH OF PEOPLE WITH CARDIO-RESPIRATORY CONDITIONS

Authors	Key features (*Additional reviewers' comments in italics*)

Table 6a-ii: Cardiac conditions - management *continued*

(NHS CRD 1998) Guidance bulletin	**Summary of research evidence on the effectiveness of cardiac rehabilitation** (National Health Service Centre for reviews and Dissemination, UK) Exercise improves physical aspects of recovery at no additional risk, but as a sole intervention it is not sufficient to reduce risk factors, morbidity or mortality – yet the majority of programmes are exercise-based. Many of the problems experienced by people with heart disease are not due to physical illness but to anxiety and misconceptions about their health. RTW rates are fairly high following an acute cardiac event, but a substantial number retire early or become unemployed. Uptake and adherence can be poor – helps if doctor strongly recommends, when access is convenient, when partner/spouse involved. (*UK Effective Health Care Bulletins are based on systematic review and synthesis of research, and produced by methodologists with expert input. Limited information on RTW but strongly advocates the biopsychosocial approach.*)
(Franklin *et al.* 1998) Narrative review	**Changing paradigms and perceptions of cardiac rehabilitation** Risk stratification has emerged as the centrepiece of strategies aimed at stabilising or enhancing the clinical status of post-myocardial infarction patients, as well as vocational counselling. The objectives of contemporary cardiac rehabilitation are to increase functional capacity, decrease symptoms, stop cigarette smoking, modify lipids and lipoproteins, decrease body weight and fat stores, reduce blood pressure and improve psychosocial well-being. Vocational counselling is often under-emphasised in contemporary cardiac rehabilitation. An occupational work evaluation may hasten return to work because it reassures the worker and their primary care physician that the physical, psychological, and environmental stresses associated with the job can be safely tolerated. Coronary risk status is much more important than functional capacity in determining RTW. (*More explicit on the 'bio' aspects of rehabilitation than the psychosocial or vocational aspects, but suggests that these dimensions are implicit in contemporary rehabilitation*).

TABLE 6: THE IMPACT OF WORK ON THE HEALTH OF PEOPLE WITH CARDIO-RESPIRATORY CONDITIONS

Authors	Key features (*Additional reviewers' comments in italics*)

Table 6a-ii: Cardiac conditions - management *continued*

(DH 2000) Policy document	**National Service Framework for coronary heart disease** (Department of Health, UK) This UK Department of Health initiative concerns how the National Health Service and others can best help people who have had a cardiac event maximise their chances of leading a full life and resuming their place in the community. Coronary heart disease (CHD) exemplifies inequalities in health: unskilled working men are three times more likely to die prematurely of CHD than men in professional or managerial occupations. The wives of manual workers have nearly twice the risk compared to the wives of non-manual workers, and angina, heart attack, and stroke are all more common amongst those in manual social classes. This National Service Framework spells out standards in 7 areas: 1) reducing heart disease in the population; 2) preventing coronary heart disease in high risk patients; 3) heart attack and other acute coronary syndromes; 4) stable angina; 5) re-vascularisation; 6) heart failure; 7) cardiac rehabilitation. The standard for cardiac rehabilitation requires that NHS Trusts put in place agreed protocols/systems so that (prior to leaving hospital after suffering coronary heart disease), patients have been invited to participate in a multidisciplinary programme of secondary prevention and cardiac rehabilitation, the aim of which is to reduce their risk of subsequent cardiac problems and to promote their return to a full and normal life. (*The Framework focuses on service provision rather than vocational issues yet return to work is an implicit goal*).
(de Gaudemaris 2000)	**Return to work with cardiovascular disease and public safety** Advances in cardiovascular therapy mean that the cardiovascular function of many patients is restored to such an extent that returning to work is possible. These authors considered that, even with ideal medical treatment, many patients should not go back to work for a variety of psychological, social, or ethical reasons (e.g. concerns about the risk to others on return to work due to diminished ability to do the job). The decision whether or not to go back to work with cardiovascular disease involves not only weighing medical considerations but also the patient's psychosocial profile and factors associated with his or her job. Work can only be resumed with the cooperation of several parties, i.e. the patient's personal physician, cardiologist, and employer. Physicians are often reluctant to send their patients back to work based on their concern about prognosis. (*Very much based on health and safety guidelines in terms of purported 'risks' to public*).

TABLE 6: THE IMPACT OF WORK ON THE HEALTH OF PEOPLE WITH CARDIO-RESPIRATORY CONDITIONS

Authors	Key features (Additional reviewers' comments in italics)
Table 6a-ii: Cardiac conditions – management continued	
(Thompson & Lewin 2000) Narrative review	**Post-myocardial infarction – rehabilitation and cardiac neurosis** Cardiac rehabilitation needs to recognise and address concomitant psychological factors. 'Home coming depression' is almost universal, and some patients become preoccupied with symptoms and become trapped in a downward spiral of increasing disability – there is evidence that psychological distress following myocardial infarction (MI) is an independent risk factor for early mortality. Beliefs and perceptions about the illness are critically important – negative models of the illness reduce likelihood of returning to work, as will blaming the problem on 'stress' or 'overwork'. Accurate information and advice (preferably written) to shift beliefs in a positive direction is important, along with provision of integrated rehabilitation services.
(Mital & Mital 2002) Narrative review	**Returning coronary heart disease patients to work** RTW or re-employment at the earliest possible time should be the ultimate goal of any cardiac rehabilitation program. Existing cardiac rehabilitation programs do not achieve a reduction in lost time from work, and programs should be based on job-related elements, rather than aerobic exercises that have no semblance to real work situations – physical or cognitive. Psychosocial factors are also more closely related to RTW outcomes than medical ones. When early RTW is the patient's major goal, a more multidisciplinary team approach is needed. (This review was combined with an individual study in which a job simulation program was devised and tested - only conclusions from review extracted).
(Wozniak & Kittner 2002) Systematic review	**Return to work after ischaemic stroke** Despite the economic cost of lost employment, return to work after ischaemic stroke has received little study. The percentages of patients working after stroke vary widely from 11 to 85%. Comparisons of these studies are difficult because they report return to work in different populations after diverse follow-up periods using variable definitions of stroke and work. Stroke severity as measured by activities of daily living was the most robust predictor of return to work. However, many factors known to influence vocational outcome after other illness (e.g., social and job characteristics) have not been examined.

TABLE 6: THE IMPACT OF WORK ON THE HEALTH OF PEOPLE WITH CARDIO-RESPIRATORY CONDITIONS

Authors	Key features (*Additional reviewers' comments in italics*)

Table 6a-ii: Cardiac conditions - management *continued*

Authors	Key features
(Womack 2003) Narrative review	**Cardiac rehabilitation secondary prevention programmes** The philosophy of early cardiac rehabilitation represents a shift in previous thinking, whereby patients did not begin rehabilitation until 6 weeks after their event. This shift is due to increased awareness of the safety of cardiac rehabilitation, and is partly related to enhanced services provided regarding secondary prevention, and partly related to the fact that many cardiac patients return to work within 2-6 weeks of hospital discharge. Patients should be encouraged to remain in medically supervised programs for longer periods or indefinitely in order that ongoing exercise supervision, careful monitoring of disease progression, and social and emotional support is provided. Cardiac rehabilitation should be population specific – depending on type of heart disease.
(Reynolds *et al.* 2004) Systematic review	**The economic burden of chronic angina** Chronic angina carries an economic burden because of symptom management, the risk of major cardiovascular events, and lost productivity. 17 studies assessed the healthcare cost of managing chronic angina - estimates varied widely because of differing patient populations, healthcare settings, countries of origin, and year(s) of data collection. 20 studies reported work limitations, 5 of which quantified productivity loss in monetary terms. Interventions for chronic angina (including revascularisation) resulted in some improvement in employment and work limitations over the short term. Chronic angina carries substantial healthcare costs caused by frequent medical visits, medications, and expensive revascularisation procedures. Workplace productivity loss because of angina is also substantial, and lasting long-term improvement in work status has been difficult to achieve. (*Does not include specific evidence on health impact of work*).
(Perk & Alexanderson 2004) Systematic review	**Sick leave due to coronary artery disease or stroke** Assessed studies of sufficient scientific quality that described sick leave following stroke, myocardial infarction, coronary artery bypass grafting, and percutaneous coronary intervention. There was limited evidence for the following results: after stroke, more than half the patients of working age returned to work during the first year following onset (higher rate for younger patients). After myocardial infarction most patients return to work. Return to work is more rapid after percutaneous coronary intervention than coronary artery bypass grafting but there is no long-term difference in sick leave. People at higher ages or with physically demanding jobs return to work to a lesser degree.

TABLE 6: THE IMPACT OF WORK ON THE HEALTH OF PEOPLE WITH CARDIO-RESPIRATORY CONDITIONS

Authors	Key features (*Additional reviewers' comments in italics*)
Table 6a-ii: Cardiac conditions - management *continued*	
(Mookadam & Arthur 2004) Systematic review	**Social support and morbidity and mortality after acute myocardial infarction** Social change, disorganization, and poverty have been associated with an increased risk of morbidity and mortality. One of the postulated mechanisms through which these determinants have been linked to health and illness is their relationship to social support. The health determinant, social isolation or lack of a social support network (SSN), and its effects on premature mortality after acute myocardial infarction is substantial: as a predictor of 1-year mortality, low SSN is equivalent to many of the classic risk factors, such as elevated cholesterol level, tobacco use, and hypertension. Because low social support is associated with increased mortality, neglecting the role of the SSN may diminish the possible gains accrued during acute-phase treatment. Therefore, lack of an SSN should be considered a risk factor for subsequent morbidity and mortality after a myocardial infarction and should be considered in cardiac rehabilitation programs and other prevention strategies.
(Newman 2004) Narrative review	**Engaging patients in managing cardiovascular health** Psychological factors play a major part in the impact, course, and treatment of cardiovascular disease. Patients' cognitions and emotions feed into their responses to their illness and its treatments, and can affect the likelihood of attendance at cardiac rehabilitation programmes. It is important to view the rehabilitation process from the perspective of the patient, to examine and assess patients' beliefs, and encourage self-management. Brief psychological interventions focused on cognitions and self-efficacy can improve the likelihood of return to work. Depression and anxiety are common after myocardial infarction and can influence outcome.
(Witt *et al.* 2005) Narrative review	**Barriers to cardiac rehabilitation** As survival after myocardial infarction (MI) improves, secondary prevention is becoming increasingly important. Cardiac rehabilitation (CR) is one modality for delivery of secondary prevention, whose ultimate goal is to help patients receive appropriate preventive therapies that will help them optimize health and reduce the risk of future cardiac diseases. However, participation rates in CR are less than optimal: in the United States, only 29.5% of MI survivors participated, in Japan 21% of those with acute MI, and in Australia 29% of those eligible were referred, and only 1/3 of those referred actually attended CR; moreover, there does not appear to be a trend towards increasing participation over time.

TABLE 6: THE IMPACT OF WORK ON THE HEALTH OF PEOPLE WITH CARDIO-RESPIRATORY CONDITIONS

Authors	Key features (Additional reviewers' comments in italics)

Table 6a-ii: Cardiac conditions - management continued

(van Dixhoorn & White 2005) Meta-analysis	**Relaxation therapy for ischaemic heart disease** A total of 27 studies in which patients with myocardial ischaemia were taught relaxation therapy were located. Physiological outcomes: reduction in resting heart rate, increased heart rate variability, improved exercise tolerance and increased high-density lipoprotein cholesterol were found. No effect was found on blood pressure or cholesterol. Psychological outcome: state anxiety was reduced, trait anxiety was not; depression was reduced. Cardiac effects: the frequency of occurrence of angina pectoris was reduced, the occurrence of arrhythmia and exercise induced ischaemia were reduced. Return to work was improved. Cardiac events occurred less frequently, as well as cardiac deaths. With the exception of resting heart rate, the effects were small, absent or not measured in studies in which abbreviated relaxation therapy was given. No difference was found between the effects of full or expanded relaxation therapy. It was concluded that intensive supervised relaxation practice enhances recovery from an ischaemic cardiac event and contributes to secondary prevention. It is an important ingredient of cardiac rehabilitation, in addition to exercise and psycho-education.

TABLE 6: THE IMPACT OF WORK ON THE HEALTH OF PEOPLE WITH CARDIO-RESPIRATORY CONDITIONS

Authors	Key features (*Additional reviewers' comments in italics*)
Table 6b: Respiratory conditions	
(Gibson *et al.* 2003) Systematic review	**Self-management education and regular practitioner review for adults with asthma (Cochrane review)** (*Assessment of asthma self-management programmes, when coupled with regular health practitioner review*). Self-management education reduced hospitalisations, emergency room visits, unscheduled visits to the doctor, days off work or school, nocturnal asthma, and quality of life. Measures of lung function were little changed. Education in asthma self-management coupled with regular medical review and a written action plan (to stimulate medication adherence) improves health outcomes for adults with asthma.
(Monninkhof *et al.* 2003) Systematic review	**Efficacy of chronic obstructive pulmonary disease self-management education programmes (Cochrane review)** Eight trials showed no effect across most outcomes from self-management education programmes, though there was a trend towards better quality of life in the educated patients. Days lost from work may not be an adequate outcome in chronic obstructive pulmonary disease patients because many are in the older age groups and often retired. Since, in most chronic obstructive pulmonary disease studies a minority of the patients undertake paid work, restricted activity days (days in which the normal activities are reduced by the disease) perhaps would be a better outcome. (*Perhaps this applies to other cardiorespiratory conditions, rendering the return to work literature biased towards conditions that permit/facilitate work*).
(Lacasse *et al.* 2003) Systematic review	**Pulmonary rehabilitation for chronic obstructive pulmonary disease (Cochrane review)** Rehabilitation was defined as exercise training for at least four weeks with or without education and/or psychological support. Rehabilitation relieves dyspnoea and fatigue and enhances sense of control over the condition – these improvements are moderately large, but improvement in exercise capacity was modest. (*There was no data on whether the health benefits of 'exercise' can be generalised to health benefits from 'work'*).

TABLE 6: THE IMPACT OF WORK ON THE HEALTH OF PEOPLE WITH CARDIO-RESPIRATORY CONDITIONS

Authors	Key features (Additional reviewers' comments in italics)
Table 6b: Respiratory conditions *continued*	
(Asthma UK 2004) Guidance	**Asthma at work** (Asthma UK – charity) There are 3.7 million adults with asthma in the UK. In most cases their asthma is not caused by work, but there can be things at work that make things worse, even to the point of triggering an asthma attack. 43% of people with asthma report that their condition can get in the way of them doing their job and more than 18 million working days are lost to asthma each year. There is no reason why people with asthma should not have almost any career they want. Many might involve triggers, but it is a matter of common sense whether to take such jobs. Employers have a general duty to protect from asthma triggers – if this is not possible, it might be worth considering another job. The right medication can help workability.
(Talmage & Melhorn 2005) (Hyman 2005) Physician guidance	**Working with common cardiopulmonary problems** *(American Medical Association guide book for primary care physicians and care providers to assist navigation of return to work issues, supported by science and consensus; the authors admit a firm belief that work is good for man).* Cardiopulmonary conditions are common, and the physician should think through the issues of risk, capacity and tolerance. As a general principle, the known risk factors that may have contributed to the disease formation should be assessed and modified (in addition to rehabilitation and return–to-work support). As with other body system problems, patients with cardiopulmonary disease are rarely harmed by return–to work recommendation. The considerable benefits of returning to work usually significantly outweigh the risk. *(Entries in respect of working with musculoskeletal conditions are in Table 5).*
(Tarlo & Liss 2005) Narrative review	**Prevention of occupational asthma** Occupational factors contribute to ~10% of adult-onset asthma, and occupational asthma is one of the commonest occupational lung diseases. Persistent asthma frequently occurs with significant socio-economic impacts. Whilst primary prevention (which is the preference) can be effective for some occupational exposures, secondary and tertiary prevention remain important. Medical surveillance programmes combined with hygiene measures and worker education are associated with improved outcomes; the sensitized worker is advised to completely avoid further areas of exposure to the sensitizer. For those who cannot avoid exposure, use of an air supply helmet respirator may help job retention. Overall, outcomes are best when workers have an early diagnosis soon after the onset of work-related symptoms, have mild asthma at diagnosis, and are removed early from further exposure.

TABLE 6: THE IMPACT OF WORK ON THE HEALTH OF PEOPLE WITH CARDIO-RESPIRATORY CONDITIONS

Authors	Key features (*Additional reviewers' comments in italics*)
Table 6b: Respiratory conditions *continued*	
(Malo 2005) Narrative review	**Work-related asthma and the impact on occupational health** Whilst occupational asthma is 'caused' by the workplace (an allergic process or a non-allergic irritant-induced mechanism), personal asthma also can 'worsen' at work (work-aggravated or exacerbated): the reasons, mechanisms, extent and consequences of this are unknown. The natural history of occupational asthma after removal from the exposure show that the majority of workers present symptoms and functional abnormalities – most improvement occurs in the first 2 years but continues at a slower rate – treatment may be needed in addition to exposure reduction. In principle, affected workers with occupational asthma can be cured (if the diagnosis is made early enough) with minimal impact on quality of life.
(Nicholson *et al.* 2005) Systematic review + guidelines	**Guidelines for prevention, identification, and management of occupational asthma** Management of the worker with occupational asthma: Surveillance should be performed for early identification of symptoms, including occupational rhinitis, with additional tests where appropriate. The outcome of interventions following a confirmed diagnosis of occupational asthma may depend on several factors, including the worker's age and the causative agent. Early diagnosis and early avoidance of further exposure, either by relocation of the worker or substitution of the hazard, offer the best chance of complete recovery. If these are impossible, workers should be relocated to low or occasional exposure areas and have increased health surveillance. Studies investigating the effectiveness of respiratory protective equipment in those with occupational asthma are limited to small studies in provocation chambers or limited case reports. The risk of unemployment may, or may not, be higher than in other adult asthmatics and may fall with increasing time from diagnosis. No studies make direct comparisons between rehabilitation systems.

TABLE 6: THE IMPACT OF WORK ON THE HEALTH OF PEOPLE WITH CARDIO-RESPIRATORY CONDITIONS

Authors	Key features (*Additional reviewers' comments in italics*)

Table 6b: Respiratory conditions *continued*

(IIAC 2006) UK Legislation	**Prescribed diseases** (Industrial Injuries Advisory Council) The UK law provides for payment of benefits to people who are suffering from certain diseases contracted in the course of certain types of employment. These diseases are referred to as prescribed diseases and are listed in Regulations. A disease can only be prescribed if the risk to workers in a certain occupation is substantially greater than the risk to the general population, and the link between the disease and the occupation can be established in each individual case or presumed with reasonable certainty. In diseases which occur in the general population (e.g. chronic bronchitis and emphysema) there may be no difference in the pathology or clinical features to distinguish an occupational from a non-occupational cause. In order to recommend prescription, IIAC looks for consistent evidence that the risk of developing the disease is more than doubled in a given occupation. In addition to a number of serious diseases (e.g. lung cancer, pneumoconiosis) there are also a few respiratory conditions that are considered prescribed diseases. These include extrinsic allergic rhinitis, emphysema, allergic rhinitis, asthma, and chronic bronchitis, all of which are associated with occupational exposure to specific agents in certain jobs. (*Whilst it is recognised that certain exposures in certain jobs are related to certain musculoskeletal diseases, it is not implied as inevitable that exposure to the job will result in the disease*) (*Also in Table 5*).
(HSE 2006) Guidance	**Asthma** Occupational asthma is the most frequently reported occupational respiratory disease in Great Britain – work-related asthma (made worse by work) is substantially more common. Key messages: • Prevent exposure, or where this is not possible keep exposures as low as is reasonably practicable • Almost all cases of occupational asthma can be prevented by use of adequate controls • Health surveillance is important. Early removal from exposure can lead to a complete recovery • Once someone has occupational asthma, low levels of exposure can provoke an attack • Once the airways become hypersensitive the disease is irreversible • Symptoms can appear immediately after exposure, or several hours later, so any link to work activities may not be obvious (*There is no suggestion that people with asthma should not work, or that work is detrimental (if suitably controlled)*).

TABLE 7: HEALTH AFTER MOVING OFF SOCIAL SECURITY BENEFITS

Table 7: Health after moving off social security benefits

Authors	Key features (*Additional reviewers' comments in italics*)
(Daniel 1983) Policy study UK	**How the unemployed fare after they find new jobs** The original Beveridge proposals assumed that unemployment was usually transient and that most people would rapidly return to work. However, the reality is that for a substantial proportion unemployment becomes long-term and is really a step to exiting the labour market. Follow-up of unemployed people who found work showed that only about 1/3rd remained in stable employment the same job, another 1/3rd were working but their employment had been unstable, and 1/3rd were unemployed again. What happened to people when they lost their job depended on age, gender, skill and occupational level. (*No data on health outcomes*).
(Moylan *et al.* 1984) Cohort study UK	**Cohort study of unemployed men** (Department of Health and Social Security) Cohort of 2,300 men who started unemployment benefit in 1978, followed up at 1, 4 & 12 months. Prior to unemployment, these men's earnings were well below the national average; few had substantial savings. Total income out of work was <50% of previous total income in work for one third of men. On return to work, 97% had an increase in income; about equal numbers had higher or lower earnings compared with before unemployment. (*This survey considered the impact of disability and sickness on unemployment and the health impact of unemployment but did not reach any clear conclusions. It did not provide any direct evidence about the health impact of coming off benefits*).
(Branthwaite & Garcia 1985) Psychological study UK	**Depression in the young unemployed and those on Youth Opportunities Schemes** (*Cross-sectional study of young people aged 16-18, n = 46*) Qualitative findings from interview showed that individual's perceptions varied with the nature of the Youth Opportunities Schemes (YOP). Those in placement regarded it as 'work' rather than training, were more satisfied and it was better for their self-esteem. Those on a Project Scheme regarded it as 'training' rather than work and few beneficial health effects were recorded. Those in employment had significantly lower scores on the Beck Depression Inventory and lower (though not significantly) scores on the neuroticism scale of the Eysenck Personality Inventory compared with those who were unemployed. The scores of those on either kind of YOP were very similar to those who were unemployed.

TABLE 7: HEALTH AFTER MOVING OFF SOCIAL SECURITY BENEFITS

Authors	Key features (*Additional reviewers' comments in italics*)

Table 7: Health after moving off social security benefits *continued*

Authors	Key features
(Svensson 1987) Brief policy paper *WHO*	**World Health Organisation perspective on social and health policies to prevent ill health in the unemployed** There is a vicious circle of poverty; unemployment, other socio-economic symptoms of inequality, (*multiple disadvantage*), vulnerability and discrimination, and chronic illness. The socio-economically disadvantaged are less likely to attain health, while the chronically ill are less likely to fulfil their socio-economic roles and may thereby become impoverished. Various policy options are discussed: to reduce unemployment; to reduce the economic consequences of unemployment and hence the health consequences of poverty; or health care to reduce the health consequences of unemployment. A holistic approach is most likely to be effective, but that depends first on awareness and consideration of the health and social consequences of unemployment. (*No evidence is presented on the feasibility, effectiveness or health outcomes of these various policies*).
(Bound 1989) Econometric study US	**The health and earnings of rejected disability insurance applicants** (Based on 1972 and 1978 Disability Surveys) Of male applicants aged >45 years for Social Security Disability Benefits who were rejected on medical grounds, <1/3rd were working at the time of survey and <50% had worked during the previous year. Rejected applicants who worked in 1977 had a mean family income of $13472 while those who did not work had $8272, compared with beneficiaries who had $10737. >50% reported that their health continued to prevent them working at all and >90% that it limited the kind and amount of work they could do. Concludes that these findings cast doubt on cross-sectional studies which suggest that the disincentive effects of DI are substantial.

TABLE 7: HEALTH AFTER MOVING OFF SOCIAL SECURITY BENEFITS

Authors	Key features (*Additional reviewers' comments in italics*)

Table 7: Health after moving off social security benefits *continued*

Authors	Key features (*Additional reviewers' comments in italics*)
(Caplan *et al.* 1989) (Vinokur *et al.* 1991b), (Vinokur *et al.* 1991a) (van Ryn & Vinokur 1992) Randomized controlled trial US	**A randomized field experiment ("JOBS") in coping with job loss** 928 recently unemployed adults were randomised to training in job seeking, problem solving, and reinforcement, on the basis that improved motivation and skills would increase job-seeking behaviour and outcomes. (*There was 59% drop-out but intention to treat analysis.*) The experimental group had significantly higher rates of re-employment than the control group (33% vs. 26% at 4 weeks and 59% vs. 51% at 4 months) They were more likely to have returned to their main occupation and had higher monthly earnings. Analysis of treatment mediation effects supported the theory of planned behaviour and the mediational role of job-search self-efficacy. There were trends for the experimental group to have lower levels of anxiety and depression and higher quality of working life, but these were non-significant. However, those who actually participated and received the intervention had effect sizes 2-3X greater on employment outcomes (e.g. earnings) and mental health (anxiety and depression). Those who were re-employed scored significantly lower on anxiety, depression and anger and higher on self-esteem, quality of life and quality of employment. Follow-up at ¹/₂ years showed that the intervention group had continued beneficial effects on level of employment, monthly income and episodes of employer and job changes. Cost-benefit analysis demonstrated large net benefits to the participants and to the federal and state government programmes that funded the project.
(Garman *et al.* 1992) Cohort study UK	**Incomes in and out of work** (Department of Social Security) Longitudinal cohort study of 3003 newly unemployed men and women in Spring 1987, with follow-up 9 months later (with 71% follow-up). Income (mainly unemployment-related benefits) and total family income in unemployment was on average less than previous earnings, though the greater the family responsibilities the less the fall. Overall, income tended to increase slightly during the course of unemployment, particularly for those with lower income to start with. For 97% of those who returned to work, net earnings exceeded their unemployment-related benefits. Allowing for all other sources of income, 63% of men and 36% of women had a total income on return tow work that was at least twice as high as total income out of work. Only 5-6% returned to work for incomes equal to or less that their total income out of work. (*This study did not provide any direct evidence on health effects, but note the strong link between social status, poverty and health inequalities*).

TABLE 7: HEALTH AFTER MOVING OFF SOCIAL SECURITY BENEFITS

Authors	Key features (*Additional reviewers' comments in italics*)

Table 7: Health after moving off social security benefits *continued*

Authors	Key features
(Erens & Ghate 1993) Cohort study UK	**Invalidity Benefit: a longitudinal survey of new recipients** (Department of Social Security) Representative national sample (n = 1545) interviewed within 1 year of starting IVB and 1-year later. At follow-up 525 remained 'attached' to the labour market – working, looking for work or still expected to return to work in future, and 535 were 'unattached' – remained unable to work and did not expect to work again. Of the attached, 19% felt their health condition had improved over past year, 31% remained the same and 27% worse; corresponding figures for unattached were 4%, 37% and 53%. About 1/5 had left IVB by follow-up and 43% of them had no continuing health problem that affected the sort of work they could do. (*These data could reflect a health selection effect and there was no evidence on a causal relationship between working/attachment to labour market and health*).
(Fordyce 1995) Task force report North America	**Back pain in the workplace** The key policy recommendation was that non-specific low back pain should recover sufficient to return to work and that there is no biological basis for long-term incapacity. Persisting disability beyond 6 weeks should therefore be re-conceptualised as a problem of activity intolerance, not a medical problem. Emphasised early intervention, re-activation and worksite-based interventions. Those who fail to restore function and return to work should be reclassified as unemployed, their disability benefits (in a workers compensation setting) should be stopped and resources put into vocational rehabilitation programs. (*The basic tenet was that work is therapeutic and improves clinical and social outcomes. These proposals created fierce international outrage and opposition*). (*Also in Table 5*).

TABLE 7: HEALTH AFTER MOVING OFF SOCIAL SECURITY BENEFITS

Table 7: Health after moving off social security benefits *continued*

Authors	Key features (*Additional reviewers' comments in italics*)
(Waddell 2004a) Narrative review	**Compensation for chronic pain** The Workers Compensation Boards of Canada are the only agencies that have attempted and the WCB of Nova Scotia (WBSNS) is the only one that actually did implement these social policy proposals in legislation and practice. Improved clinical management for the (prevention of) chronic pain, early intervention, rehabilitation, and support into work interventions were generally welcomed. The attempt to time-limit benefits led to extensive political debate, which in most provinces blocked any legislation. The legislation in Nova Scotia led to legal challenge that was taken all the way to the Supreme Court of Canada which effectively ruled that any attempt to set an arbitrary time-limit was in breach of human and constitutional rights of access to services and benefits, and discriminatory. (*Unpublished administrative data from WBSNS showed that the package of case management, rehabilitation and benefit changes led to improved clinical outcomes and a higher proportion of claimants returning to work (G Waddell, personal communication). However, there is no direct evidence that return to work per se improved health outcomes*).
(Vinokur *et al.* 1995) (Vinokur & Schul 1997)	**Impact of the JOBS intervention on unemployed workers varying in risk of depression** (*This is a separate cohort from Caplan et al 1989.*) 1801 recently unemployed adults, 671 of whom participated in a modified JOBSII intervention, which focused on enhancing their sense of mastery through the acquisition of job-search and problem-solving skills and on inoculation against setbacks. Those receiving the experimental intervention had significantly greater improvement in the primary outcome of their 'mastery' score, both as a Condition main effect and a Condition X Time interaction effect. A significantly higher proportion of the experimental group were re-employed at follow-up; this effect was strongest in those with higher baseline depression scores. The experimental group also had a significant and significantly greater improvement in depression scores, and this effect was again strongest in those with higher baseline depression scores. There was evidence that both re-employment outcomes and mental health outcomes had a reciprocal relationship, each one influencing the other over time. Modelling showed that mastery and inoculation against setbacks mediated the impact of the programme. (*There was a high drop-out rate but intention to treat analysis*). (continued)

TABLE 7: HEALTH AFTER MOVING OFF SOCIAL SECURITY BENEFITS

Authors	Key features (*Additional reviewers' comments in italics*)

Table 7: Health after moving off social security benefits *continued*

Authors	Key features
(Vinokur *et al.* 2000) Randomized controlled trial US	**Impact of the JOBS intervention on unemployed workers varying in risk of depression** (continued) Two-year follow-up (Vinokur *et al* 2000) showed that the experimental group still had significantly higher rates of employment and monthly income, lower levels of depressive symptoms, fewer major depressive episodes, and better role and emotional functioning than the control group. Those with lower baseline levels of motivation and mastery had greater benefits on both employment and mental health outcomes. Baseline depression and financial strain were risk factors while job-search motivation and sense of mastery were protective for mental health outcomes. Analysis of the interactions showed that at least part of the intervention effects on mental health were mediated by re-employment. In contrast to JOBS (Caplan *et al* 1989), the JOBSII intervention did not demonstrate any improvement in quality of work, possibly due to the different economic situation. (*Complex statistical modelling*).
(Rowlingson & Berthoud 1996) Cohort study UK	**Longitudinal cohort study of Disability Working Allowance recipients** (Department of Social Security) Two cohorts, n = 886 + 629 with 1- and 2-year follow-up respectively. Those with less severe disabilities and those who experienced an improvement in their health condition were more likely to move into work. (*Because of the limitations of the available data*) it was difficult to know whether the improvement occurred before or after movement into work, so it was not possible to draw any firm conclusions about cause and effect. Those surveyed considered working preferable to being on benefits because it provided a social identity, meaning and interesting activity, social contact and financial independence from the state. They felt the main attractions of particular jobs were good wages, job security and food relations with employers. Few people managed to move into work, but for those who did it was when their health condition improved, when they found a job that accommodated their condition or when they had a sympathetic employer.

TABLE 7: HEALTH AFTER MOVING OFF SOCIAL SECURITY BENEFITS

Table 7: Health after moving off social security benefits *continued*

Authors	Key features (*Additional reviewers' comments in italics*)
(Proudfoot *et al.* 1997) Randomised controlled trial UK	**Effect of cognitive-behavioural training and job-finding among long-term unemployed people** 289 professional people (social classes 1, 2, 3 & 7) who were unemployed >12 months were randomised to CBT or a control group. There was 31% drop-out and 8% loss to follow-up. Follow-up was at 4 months after completing the programme. Training produced significant improvements in self-esteem, self-efficacy, GHQ, life satisfaction and motivation for work, but many of these decayed by follow-up. The experimental group had significantly more in full-time work (34% vs. 13% $P<0.001$) and significantly greater improvement ($P<0.05$) in GHQ at 4-month follow-up. When those in full-time work were excluded from the analysis there were no significant group differences, so it appears that the improvement in mental health only persisted in those who had returned to work. (*There was no further sub-group analysis of the link between return to full-time work and improvement in mental health*).
(Dorsett *et al.* 1998) Cohort study UK	**Leaving Incapacity Benefit** (Department of Social Security) Longitudinal cohort study of 2263 people leaving IB between June and November 1996, interviewed 5-10 months after leaving IB, postal follow-up 6-9 months later with 63% response rate. 36% left IB voluntarily, 64% because they were disallowed i.e. 'failed' the DSS medical test of incapacity. 32% returned to full or part-time work, 29% were unemployed and seeking work, and 26% remained 'sick' (and 13% miscellaneous). By the time of follow-up, 38% remained in full- or part-time work, 27% had returned to IB and 22% were not working and on other social security benefits (and 13% miscellaneous). Net weekly income at the end of their IB claim was £171 and at interview 5-10 months later was £177. However, there were two sharply divergent courses of leaving IB. Of those who left IB voluntarily, 68% returned to some form of work and their income at follow-up rose to £234. Of those who were disallowed IB, only 19% returned to work and their income at follow-up ranged from £130-152. Of those who appealed the decision, only 4% returned to work, and winning or losing their appeal made little difference. Overall, 88% reported continuing health problems on leaving IB but there were important differences by route of leaving. At the time of leaving benefits, 55% of voluntary leavers felt their health condition was fully recovered or much better, compared with 13% of those who were disallowed. By the time of interview, 47% of those who left benefits voluntarily felt that their health had fully recovered or was much better, compared with 24% of those disallowed who did not appeal and 4% of those who were disallowed and appealed. At follow-up, the corresponding figures were 55%, 37% and 9%, while the proportion saying their health had got worse was 13%, 20% and 50%. Overall, three-quarters said they had some continuing work restrictions. (*These findings are all self-reported perceptions of health. There are methodological problems interpreting how people rate change in health status retrospectively over time. The health data was not analysed in relation to return to work*).

TABLE 7: HEALTH AFTER MOVING OFF SOCIAL SECURITY BENEFITS

Authors	Key features (*Additional reviewers' comments in italics*)

Table 7: Health after moving off social security benefits *continued*

Authors	Key features
(Kornfeld & Rupp 2000) Social security RCT US	**The net effects of Project Network** Project Network was an RCT of an intensive outreach, case management and work incentive package for 8248 SSDI and SSI applicants and recipients in 1991. The intervention produced statistically significant increases in average earnings in the two years subsequent to the intervention, but only by 11% and insufficient to make any substantive improvement in the living standards of the participants. There was negligible effect on total benefits paid. There was no significant effect on self-assessed health or improvement in health, functional limitations, work limitations, emotional problems, Mini Mental State Examination or Mental Health Inventory. (*The health effects were not analysed separately for those who did/did not return to work*).
(Rosenheck *et al.* 2000) Cohort study US	**Outcomes after initial receipt of social security benefits among homeless veterans with mental illness** Homeless veterans with serious mental illness (n = 50) who received benefits did not differ on any sociodemographic or clinical measure at baseline from those who were denied benefits (n = 123). At 3-month follow up, beneficiaries had significantly higher total income and reported higher quality of life, but there was no significant difference on standardised measures of psychiatric status or substance abuse, or in attitudes to work.

TABLE 7: HEALTH AFTER MOVING OFF SOCIAL SECURITY BENEFITS

Authors	Key features (*Additional reviewers' comments in italics*)

Table 7: Health after moving off social security benefits *continued*

(Ford *et al.* 2000) Conceptual narrative review UK	**Welfare to work: the role of general practice** Explores the relationship between unemployment, ill health and GP sick certification. Social security and employment policy initiatives are discussed in relation to the literature on the association between unemployment and ill health, sociological and psychological perspectives on work and unemployment, medicalisation of unemployment, adjudication of fitness for work, re-employment and health, and barriers to (re)-employment. The adverse effects of unemployment on health, particularly mental health, include a raised standardised mortality rate from suicide and accidents, increased prevalence of depression, anxiety and somatic illness, and increased GP consultation rates. However, the relationship between unemployment and ill health is complex, combining elements of both cause and effect. Factors moderating the impact of unemployment on health include financial status, social support, age and duration of unemployment, previous job satisfaction, expectancy of re-employment and perceived responsibility for job loss. Most of the adverse health effects are reversed by re-employment. In the majority of cases, re-employment and ensuing economic independence will be as beneficial to the individual and his or her family as it is to the taxpayer. Welfare to Work policy is likely to improve health if it helps jobless individuals to find meaning and purpose in paid employment. However, there may be many barriers to re-employment such as successful adaptation to life on benefits, when financial gain from re-employment may be very small compared with the satisfaction of the person's present lifestyle, e.g. looking after a family (*or in retirement*). The long-term unemployed may, over time, become virtually unemployable. Inappropriate pressure on those frail in body or mind, or who have adapted well to a life dependent on benefits may be harmful to some individuals. GPs play a key role with sick certification and advice on fitness for work. GPs also bear responsibility for mobilising resources to improve their patients' health, while protecting them against unreasonable and unrealistic expectations.

TABLE 7: HEALTH AFTER MOVING OFF SOCIAL SECURITY BENEFITS

Authors	Key features (*Additional reviewers' comments in italics*)

Table 7: Health after moving off social security benefits *continued*

Authors	Key features
(Ashworth *et al.* 2001) Research Report UK	**Well enough to work?** (Department of Social Security) Secondary data analysis of the Leaving Incapacity Benefit (IB) survey (Dorsett et al 1998) and the Jobseekers' Allowance (JSA) survey. 30% of JSA recipients reported work-limiting problems with their health and there was considerable movement between JSA. 5% of recipients leaving JSA moved on to IB within 6-9 months. 24% of IB leavers moved onto JSA within 1-month. The vast majority of entrants to JSA from IB had been disallowed IB because they failed the DSS medical test of incapacity. Only 9% of them said their health had improved, 38% that it was the same, and 23% worse than when they started IB. By follow-up, 61% of IB leavers who had returned to work said their health had improved, compared to 41% of those who had moved to JSA; 80% of those working said their health continued to affect the kind of work they could do, compared to 94% of those who remained on JSA. Movers (either from JSA to IB or vice versa) had more multiple disadvantages in the labour market than other JSA or IB recipients, even apart from their health condition, (*and it is not clear from this data how much these movements between benefits was due to the health condition, a health selection effect, or had any impact on health*).
(Hedges & Sykes 2001) Research Report UK	**Qualitative study of movers between JSA and IB or Income Support for sickness (ISS)** (Department for Work and Pensions) Moves from JSA to IB were usually caused by the onset, recurrence or deterioration of a health problem and were generally seen to be appropriate. Moves from IB/ISS to JSA usually resulted from doctors stopping sick certification or failing the DWP medical test for incapacity (PCA). In these cases some claimants did not agree they were now well enough to work. However some others decided for themselves that they were fit enough to go back to work without being told to. (*No data on health impact of these moves*).

TABLE 7: HEALTH AFTER MOVING OFF SOCIAL SECURITY BENEFITS

Authors	Key features (*Additional reviewers' comments in italics*)

Table 7: Health after moving off social security benefits *continued*

Authors	Key features
(Bloch & Prins 2001) Social security study International	**Who returns to work and why?** International Social Security Association longitudinal study of cohorts each of 3-600 incapacity benefit recipients with incapacitating LBP >3 months in Denmark, Germany, Israel, the Netherlands, Sweden and the United States with 2-year follow-up. Aimed to investigate the timing and effects of medical and rehabilitation interventions to restore health and work capacity. The less severe the pain and the better the functional capacity, the higher the work resumption rates in all cohorts. (*This was generally presented as a health selection effect in return to work.*) However, there were marked differences in resumption levels within equal levels of pain intensity and functional capacity with high cross-national differences. In most cases, medical treatment resulted in an improvement of subjective health status, but this did not generally correspond to improvement in back function and pain intensity or influence the chances of return to work. Overall, neither the health condition nor medical turned out to be absolute indicators of the probability of work resumption, and the study concluded that the whole question of the relationship between health status and long-term (in)capacity for work needed further research. Conversely, it was noted that subject who had resumed work within one year had statistically better back function at two years than those who did not. General health in Sweden, Germany and the US and mental health and social functioning in all countries was also better at two years among those who had resumed work within one year than amongst those who had not.
(Vuori *et al.* 2002) Randomized controlled trial Finland	**The Tyophon job search programme in Finland** Replication of the JOBS (Caplan et al 1989) and JOBSII (Vinokur et al 1995) programs in Michigan, USA. 1261 unemployed job-seekers (ranging from 3 months before lay-off to 5 years of unemployment) who volunteered to participate, with 6 month follow-up. The programme was designed to use an active learning process to boost job-search motivation and enhance job-search skills. 34% of the experimental group and 32% of the control group were re-employed at follow-up (ns) though re-employment in the experimental group was significantly more likely to be sustained. The intervention had no significant effect on psychological distress, depressive symptoms or job satisfaction. It is suggested that these very different results from similar programmes in US could be due to the different subjects (many with longer duration of unemployment), major labour market differences and different social and social security contexts. In particular, the US unemployed receive lower levels of financial support for shorter duration, which may increase levels of psychological distress and enable the intervention to have greater effects on mental health outcomes. It is also unclear whether the intervention would have worked on US workers who had been unemployed >3 months.

TABLE 7: HEALTH AFTER MOVING OFF SOCIAL SECURITY BENEFITS	
Authors	**Key features** (*Additional reviewers' comments in italics*)

Table 7: Health after moving off social security benefits *continued*

Authors	Key features
(Juvonen-Posti *et al.* 2002) Quasi-experimental study Finland	**Return-to-work rehabilitation and re-employment project** 140 middle-aged (age >35 years), long-term unemployed with various disabilities participated in a 'Pathway-to-Work' Project with a matched control group and 6 months follow-up. The project combined treatment in a rehabilitation unit with service provided by the local employment office and the municipal authorities. Clients received individual work-oriented guidance, counselling, a group support programme, tailored vocational rehabilitation, work tryouts at workplaces and subsidized placement, all coordinated by case management. (*This programme appears to bear many similarities to the UK Pathways to Work pilots (Waddell & Aylward 2005)*). Those receiving the experimental intervention had higher participation rates than the control group. At the end of the project, 48% of the intervention group were working compared with 9% of the control group (P < 0.001). 86% of the control group remained unemployed compared with 16% of the experimental group. However, 21% of the experimental group were now on sick leave or retired compared with 2% of the control group (P <0.001) and most of them then remained sick over the next two years. Psychological distress (measured on the GHQ) improved significantly in the experimental group but mental health data for the control group was not presented for comparison. (*These results do not permit any conclusions about whether the experimental intervention or re-employment had any impact on mental health outcomes*).
(Scheel *et al.* 2002c), (Scheel *et al.* 2002a), (Scheel *et al.* 2002b) Cluster-randomised controlled trial Norway	**Active Sick Leave for social security claimants with back pain** Active Sick Leave (ASL) was a social security benefit scheme offered to temporarily disabled workers to promote and support early return to modified work. 6176 claimants with >16 days sick leave due to back pain were randomised by municipality (n=65) to a proactive intervention, a passive intervention or a control group. Those in the proactive group were more likely (17.7% vs. 11.5%, P=0.02) to use the ASL arrangements. There was no significant difference between the groups in average duration of sick leave, proportion returned to work at 1-year, recurrent sick leave for back pain, or the proportion going on to long-term disability. Self-report questionnaires showed no significant difference between the groups in bodily pain or physical functioning (SF-36), quality of life or satisfaction with various aspects of management. (*However, the validity of the latter analysis was limited by the low take-up of ASL and a 38% return rate of the questionnaires and no conclusions can be drawn about any impact of the intervention on quality of life.*). The minority who actually participated in ASL had significantly shorter sick leave and returned to work sooner, though it is not possible to say if this was due to the intervention or a self-selection effect. (*The authors point out that the participants in ASL did not receive any work-site support*).

TABLE 7: HEALTH AFTER MOVING OFF SOCIAL SECURITY BENEFITS

Authors	Key features (*Additional reviewers' comments in italics*)

Table 7: Health after moving off social security benefits *continued*

(Bacon 2002) Statistical analysis UK	**Moving between sickness and unemployment** (Department for Work and Pensions) Substantial numbers of social security benefit clients move between Jobseeker's Allowance (JSA) and Incapacity Benefit (IB). From 1999-2000 data, about 190,000 clients left JSA and started IB within a month and about 110,000 moved from IB to JSA each year. Overall, a quarter of those leaving IB moved to JSA: of those who left IB voluntarily, 10% claimed JSA within a month and 69% were employed; of those disallowed IB, 32% claimed JSA within a month and only 10% were employed. Ex-IB recipients were then more likely than any other group to remain on JSA and least likely to (re-)enter work – only 25% within 6-9 months. Those who experienced both unemployment and ill health or disability (i.e. had spells on both JSA and IB) tended to share multiple disadvantages in the labour market: many had been out of the labour market >2 years; they were more likely to be older, less well qualified, have problems with basic skills, and not to have access to private transport. Those who moved from JSA to IB were more disadvantaged than those who moved from IB to JSA. Many movers said they would like to work and thought that work had many benefits: helping to avoid a sense of dependency, depression and isolation; restoring self-respect, confidence, a sense of personal worth and independence; and providing a means of keeping active. However, 89% had some degree of continuing health problems: fluctuating and enduring health problems were most difficult to combine with work. They lacked confidence in their ability to find a job they would be able to do, get that job, do the job without aggravating their health problems, or keep the job. 51% of those who moved from IB to JSA to economic activity said their health was improved to some extent; compared to 31% of those who then remained on JSA. (*However, there is likely to be a health selection effect and there is no direct evidence of a causal link between (re-)entering work and health improvement*).

TABLE 7: HEALTH AFTER MOVING OFF SOCIAL SECURITY BENEFITS

Authors	Key features (Additional reviewers' comments in italics)

Table 7: Health after moving off social security benefits *continued*

Authors	Key features (Additional reviewers' comments in italics)
(Corden & Thornton 2002) In-House Report International comparison	**Employment programmes for disabled people: lessons from research evaluations** (Department for Work and Pensions) Evaluation of 6 government programmes in UK, USA, Canada, Australia and Austria, supplemented by a literature review. Those who participated and completed the programmes generally had a higher rate of return to work (though this may reflect motivation and health selection effects). Programme participants reported both positive and negative effects of programme participation on their general health and well-being. Project NetWork (USA) was the only study which included a controlled impact analysis on health and well-being, and generally did not show statistically significant effects on various measures of health and well-being. Some participants in the New Deal for Disabled People (UK) reported positive impacts such as improved self-esteem, confidence and sense of security, but others reported negative impacts such as increased stress, disappointment and loss of self-esteem, especially when the pace of progress appeared slow. (*These results were all reviewed in terms of the impact of the programmes, not in relation to whether or not participants actually returned to work*).
(Smeaton & McKay 2003) Research Report UK	**Working after State Pension Age** (Department for Work and Pensions) Secondary analysis of three large national databases. There was a strong link between people's perceptions of their health, and the likelihood that they were in paid work after State Pension Age: 15% of men and 14% of women with 'excellent' health were in paid work, but no-one who described their health as 'very poor' was working. Conversely, 76% of men who were working aged 65-75 described their health over the last years as either 'excellent' or 'very good', compared with 54 per cent of non-workers. Among women, 71% of workers reported a similarly high level of health, compared with 49% of non-workers. Those working also had higher scores on the GHQ. More detailed analysis attempted to control for previous health and employment experience. These did not show any clear health selection effect (though numbers in some samples were small). This suggested that the ability of people to sustain or even improve health was better among continuing workers than among those remaining out of the labour market.

TABLE 7: HEALTH AFTER MOVING OFF SOCIAL SECURITY BENEFITS

Authors | **Key features** (*Additional reviewers' comments in italics*)

Table 7: Health after moving off social security benefits *continued*

Authors	Key features (*Additional reviewers' comments in italics*)
(Bowling *et al.* 2004) In-House Report	**Destination of Benefit Leavers** (Department for Work and Pensions) Survey of benefit recipients leaving various benefits in February – April 2003. 49% of IB leavers entered work of 16+ hours/week which is significantly higher than previous estimates. 21% moved on to another benefit for people who are out of work (3/4 of them to JSA).
(Coleman & Kennedy 2005) Research Report UK	Further survey of 17166 benefit leavers in February – May 2004. 50% of IB leavers entered work of 16+ hours/week. 19% moved on to another benefit, which is the highest of all client groups. (*No data on health outcomes*)

(Leech 2004)
Intervention in social security setting
Ireland
	Preventing chronic disability from low back pain
	(Department of Social and Family Affairs)
	A study in a social security setting that aimed to determine if early intervention, using international evidence-based guidelines in the assessment of claimants with back pain, would decrease the incidence of progression to chronic disability. Compared results of the programme with historical data. Medical assessors were trained in evidence-based practice and the project was promoted at the professional level and patient information was made available.
	
	• As expected 52% returned to work within 4 weeks of their own volition
	• Approximately 1600 claimants selected to attend for medical assessment at 4 to 6 weeks from date of claim (this being much earlier than previously)
	• On receipt of the invitation, 63% came off benefits and returned to work
	• The remainder were duly assessed:
	  • 64% of LBP cases declared fit for work compared with ~20% previously
	  • There were fewer appeals and fewer successful appeals
	  • There was reduced duration of claim and reduced benefit costs
	
	The report concluded that this early intervention in the acute stage should result not only in the improved health of back pain patients, but also in decreased health care costs, reduced absenteeism, increased production, and significant savings in long-term illness benefit schemes. (*But, these data were not collected during the 6-month project*).

TABLE 7: HEALTH AFTER MOVING OFF SOCIAL SECURITY BENEFITS

Authors	Key features (*Additional reviewers' comments in italics*)

Table 7: Health after moving off social security benefits *continued*

Authors	Key features
(Watson *et al.* 2004) Pilot study UK	**Back to Work** (Department for Work and Pensions) Cohort study of a work-focused rehabilitation programme for social security benefit recipients with chronic low back pain (n = 88). 91% of those who started the programme completed it, and at 6-month follow-up 71% of them were employed, on work placement or in education/training. When replicated in a second centre, 64% had comparable outcomes. Data on the Roland Disability Questionnaire (RDQ) were re-analysed (P Watson, personal communication) separately for those who did and those who did not return to work. These groups had almost identical RDQ levels at baseline and almost identical, significant improvement by the end of the intervention. Those who returned to work showed slight (non-significant) further improvement in RDQ at 6 months. Those who did not return to work showed progressive deterioration in RDQ at 3 and 6 moths. By 6 months those who had returned to work had significantly ($P < 0.05$) better RDQ levels than those who did not.
(Mowlam & Lewis 2005) Qualitative study UK	**How General Practitioners work with patients on sick leave** (Department for Work and Pensions) Qualitative study of 24 GPs to investigate their approaches to managing sickness absence and assisting patients to return to work. There was a recurrent view among GPs that work could be of therapeutic benefit to patients. A number of reasons were given. First, GPs noted a range of psycho-social benefits associated with working. The activity of working and the financial independence it brings were seen as contributing to self-esteem and self respect, making people feel valued and helping their inclusion within wider society. Having more money from being employed, and a resulting better standard of living, was also felt to impact positively on health. Being off work sick was also widely thought to pose a risk to people's mental health. This was partly explained by the lack of routine and social interaction resulting from being off work. GPs talked about people being more isolated when not working, and felt that this and having too much time on their hands often led people to dwell on things, and to become more anxious and depressed. Work was described as playing an important role in providing people with a routine as well as a social network. (*continued*)

TABLE 7: HEALTH AFTER MOVING OFF SOCIAL SECURITY BENEFITS

Authors	Key features (*Additional reviewers' comments in italics*)

Table 7: Health after moving off social security benefits *continued*

Authors	Key features
(Mowlam & Lewis 2005) Qualitative study UK	**How General Practitioners work with patients on sick leave** (*continued*) **(Department for Work and Pensions)** GPs also described how being at work could actually constitute an important part of the management of some conditions. Work was seen as helping to sustain people's physical capacities and aid their recovery, as well as keeping their work skills fresh. Even if the job was not a particularly physical one, GPs talked about the benefits of being active, for example in terms of the journey to and from work. Depending on the type of activities undertaken, remaining at work could constitute a key element of a treatment pathway for a patient. In addition, there was recognition that the longer patients were absent from work, the harder it could be for them to return. Overall, then, there was a clear perception of the potential therapeutic benefits of work – both physical and psychological. There were, however, a number of caveats. GPs felt that whether work actually was a source of self respect and personal fulfillment depended very much on the type of work being undertaken. These GPs argued that low wage or menial jobs had little social status and did not contribute anything like the same levels of self esteem that a highly respected job might, and could be bad for people's mental health in its broadest sense. In addition, some GPs were not at all sure that people were better off being in a low wage job than being on state benefits. They also highlighted that the job was, itself, sometimes the actual cause of the health problem. There were some references to physical conditions here, where work activities either caused or had exacerbated a physical condition. However, much more emphasis was given to anxiety and stress caused by work, where there were different views about whether a return to work could be therapeutic. Some GPs felt that signing people off sick under these circumstances resulted in unwelcome delay to finding a resolution to the problem. They argued that unless the person was actually there to sort it out, the problem would only be waiting for them on their return to work, which could lead to real anxieties about going back. However, another viewpoint was that some time away from the workplace was necessary if stress undermined their ability to work, and for the patient to recover. These sorts of considerations and experiences led some GPs to be less convinced of the therapeutic value of work, or to see it as less true in some circumstances.

TABLE 7: HEALTH AFTER MOVING OFF SOCIAL SECURITY BENEFITS

Authors	Key features (*Additional reviewers' comments in italics*)

Table 7: Health after moving off social security benefits *continued*

Authors	Key features
(Kazimirski *et al.* 2005) Cohort survey UK	**New Deal for Disabled People (NDDP) Survey** (Department for Work and Pensions) Survey of 4082 people on incapacity benefits who participated in NDDP in 2002, interviewed 4-5 and again 13-14 months later, by which time 60% were economically active, including 43% working. Over the course of the study, 30% felt their self-perceived general health had improved, 47% that it had remained the same and 23% that it had deteriorated. 30% reported they had less limitations in their normal everyday activities, 47% that they were the same and 21% that they were more limited. People's predictions of their likely future course at the time of first interview were not very accurate: one third of those who expected their health condition to remain stable actually changed; 45% of those who expected it to change actually remained stable. People whose health remained bad or very bad tended to have multiple disadvantages and some of them felt their Job Broker had been unhelpful regarding their health. (*There was no other analysis of any relationship between participation in NDDP or return to work and health outcomes*). Mean levels of participation in social activities, levels of satisfaction with life in general, high satisfaction with family life and low satisfaction with their financial situation remained stable over the course of the study, though again a significant minority either improved or deteriorated.

TABLE 7: HEALTH AFTER MOVING OFF SOCIAL SECURITY BENEFITS

Authors	Key features (*Additional reviewers' comments in italics*)

Table 7: Health after moving off social security benefits *continued*

Authors	Key features
(Waddell & Aylward 2005) Conceptual, narrative review UK	**The scientific and conceptual basis of incapacity benefits** Reviews the evidence on the negative impact of long-term incapacity benefits: For too long, people who did still have (some) capacity for work received no active support for rehabilitation or reintegration but were effectively written off and consigned to a lifetime on benefits. Once people are on IB, physical and mental health tends to deteriorate rather than improve. The benefit system itself creates 'welfare dependency', lowers self-esteem and denies opportunity and responsibility in almost equal measure. The barriers to coming off benefit and returning to work become greater and the chances of doing so become less over time. In summary, receipt of IB often reinforces 'incapacity' and becomes a barrier to work. Tragically, the IB regime created needless incapacity in some people whose incapacity should have been avoidable, if they had been given the right opportunities and support. Argues that work is the best exit from incapacity benefits and, in that sense, 'work is the best form of welfare'. Summarises the reasons as: 1) many disabled people do work and many more want to work; 2) the medical evidence is that many people on 'incapacity' benefits with longer-term sickness still have capacity for (some) work despite their health condition; 3) work has health-related, personal and social advantages over worklessness; 4) incapacity benefits have many financial and other disadvantages compared with the financial and other advantages of earned income – on average, disabled people in work earn 80-90% as much as non-disabled workers, while disabled people who do not work earn less than half that amount. (*Provides further discussion and references to support these arguments, but does not provide any direct evidence about the health impact of coming off incapacity benefits*).

Appendix

REVIEW METHODS

The questions posed for this review required consideration of a wide range of evidence, of varying type and quality, from a range of disciplines, methodologies, and literatures. This required a flexible approach that was allowed to develop as the project progressed. At the same time, a rigorous approach was required when it came to assessing the strength of the scientific evidence. Combining these requirements necessitated a combination of review methodologies.

Systematic reviews have become the preferred method for analysing scientific literature, and are well suited to addressing specific questions, e.g. on the effectiveness of interventions or assessing the strength of a particular set of evidence. However, that methodology is less appropriate for addressing broader questions and different kinds of evidence (Horan 2005). An alternative approach is a 'best evidence synthesis', which summarises the best available evidence and draws conclusions about the *balance of evidence,* based on its quality, quantity and consistency (Slavin 1995; Franche *et al.* 2005). Quality refers to the methodological quality of studies that can be considered acceptable; e.g. sample size, external validity, minimising bias. Quantity refers to the number of (acceptable) research studies on a particular issue. Consistency refers to the consistency of findings across studies, and represents a balance of the available information. Naturally, a primary criterion for including studies is their relevance for the purpose of the review and the research question. Importantly for the questions concerned here, a best evidence synthesis approach offers quality assurance together with the flexibility needed to tackle heterogeneous evidence and complex socio-medical issues.

In summary, the aims of this review were defined in broad general terms, searches were designed to be very inclusive, and a number of the evidence statements represent post-hoc interpretations depending on what was found in a voluminous literature. Inevitably, this makes the review difficult to reproduce and requires a measure of trust in the authors' judgement.

THE STRUCTURE OF THE EVIDENCE AND LITERATURE REVIEWED

For the purposes of this review and for literature searching, 'work', 'health' and 'well being' were interpreted broadly. There was a particular focus on the common health problems that account for most sickness absence, long-term incapacity and early retirement - minor/moderate mental health, musculoskeletal, and cardio-respiratory conditions (Waddell & Burton 2004); so major trauma and serious disease were included only if the evidence was particularly illuminating.

The basic process was to develop, in general terms, the key topics of interest and then to build up a comprehensive set of literature databases. From the starting premise, the logical entry point was reviews on the health effects of unemployment. It was immediately apparent that that evidence base is not solely about unemployment, but is actually a comparison of unemployment and working, and that any health effects may be positive or negative. Most of those reviews were of young or middle working-age adults, so a specific search was made for material on older workers, which also introduced issues around retirement.

Whilst that material covered loss of employment, it did not include any systematic review of the effects of re-employment. A separate search was therefore made for individual longitudinal studies on the health impact of (re-)employment.

Only then was it apparent that all of the evidence retrieved to that point was about the impact of work or unemployment on people who were, by definition, healthy. It did not address the question of whether or not work was good for people who were sick or disabled. An additional search was therefore made for reviews about the impact of work on the health of people who are sick or disabled; this came from a different, largely clinical literature. The retrieved material demonstrated a broad consensus of opinion but provided very little actual scientific evidence, so separate searches were made in the three key health categories: mental health, musculoskeletal, and cardio-respiratory conditions.

Finally, recognising that social security benefit recipients may represent a special case, a separate search was made for evidence on the health impact of coming off benefits and re-entering work. English language publications in this area are mainly from the UK and the US, which may not be entirely relevant to other countries. Studies from workers compensation and similar environments were excluded because the findings are not readily generalisable.

The overall shape of the review and the type of material included is shown in Table A1; the data were inserted into the evidence tables (Tables 1 to 7), in chronological order.

LITERATURE SEARCHING AND SELECTION

Standard search methodologies have major limitations for this kind of poorly defined and ill-standardised literature, which includes a variety of scientific and grey literatures that are covered to variable extent by different electronic databases, and for which there is often no standard indexing.

Search strategies were therefore broad, taking multiple and overlapping approaches to each area of interest. Electronic databases and search engines included Pubmed; psychINFO; EMBASE; CINAHL; OTseeker; JSTOR; Google Scholar; Nioshtic; ISI Web of Knowledge; SCOPUS. Depending on the topics and the databases, numerous combinations of keywords were used either singly or in combination, using Boolean operators and truncation where appropriate: e.g. work, occupation, employment, unemployment, re-employment, health, ill health, well-

TABLE A1: The key stages of the review and the material included

Stages of review	Included material	Evidence table
Health effects of work vs. unemployment	Systematic reviews, Meta-analyses, Narrative reviews, Policy papers	Table 1
Health impacts of re-employment	Longitudinal studies	Table 2
Work for sick and disabled people	Reviews, Policy statements, Guidance	Table 3
The impact of work on people with mental health conditions	Systematic reviews, Meta-analyses, Narrative reviews, Policy papers	Table 4
The impact of work on people with musculoskeletal conditions	Systematic reviews, Meta-analyses, Narrative reviews, Policy papers	Table 5
The impact of work on people with cardio-respiratory conditions	Systematic reviews, Meta-analyses, Narrative reviews, Policy papers	Table 6
Health after moving off social security benefits	Reviews, Policy papers, Longitudinal studies, Intervention trials	Table 7

being, sickness, sickness absence, return to work, retirement, and early retirement. Condition-specific keywords were added when seeking papers on mental health, musculoskeletal disorders and cardio-respiratory conditions (using MESH terms where appropriate), together with the search terms recommended for retrieving literature on chronic disease and work participation - work capacity, work disability, vocational rehabilitation, occupational health, sick leave, absenteeism, return to work, retirement, employment status, and work status (Haafkens *et al.* 2006). The electronic searches were restricted to literature reviews, except for re-employment (Table 2) where longitudinal studies were specified, and for moving off social security benefits (Table 7) where all types of study were included.

Electronic searches were supplemented by personal databases, communication with experts in the field, official reports and other grey literature. Extensive citation tracking was performed, and in some poorly indexed areas this produced a sizeable proportion of the relevant articles. Useful websites included: the UK Department of Health (**www.dh.gov.uk**); the UK Department for Work and Pensions (**www.dwp.gov.uk/** and **www.dwp.gov.uk/asd/asd5/**); the World Health Organisation (**www.who.int**); the Organisation for Economic Cooperation and Development (OECD) (**www.oecd.org**); the Council of Europe (**www.europa.eu.int**); the UK Health & Safety Executive (**www.hse.gov.uk**); the US National Institute of Occupational Safety and Health, (**http://www.niosh.com.my/**); the European Agency for Safety and Health at Work (**http://agency.osha.eu.int/OSHA**); the US Occupational Safety and Health Administration (**www.osha.gov**); the US Social Security Bulletin (**www.ssa.gov/policy/docs/ssb**); the UK Disability Rights Commission (**www.drc-gb.org**) and linked sites; the UK National Institute for Health and Clinical Excellence (NICE) (**www.nice.org.uk**); the UK Health Development Agency (HAD) (**www.publichealth.nice.org.uk**) and related sites; and the UK Work Foundation (**www.theworkfoundation.com**).

The focus was on adults of working age, generally in the range 16 to 65+ years. Only publications in English were included. Reviews published from 1990 onwards were taken to supercede earlier publications and to be most relevant to modern work. Earlier reviews (identified from citation tracking and personal databases) were only used if they were considered classical sources. Literature acquisition extended through to January-March 2006.

The range of acceptable material was broadly similar to that used previously for this kind of review (Waddell & Burton 2001; Waddell & Burton 2004; Burton *et al.* 2004). Systematic reviews, which provide the most robust scientific evidence, were used wherever possible and given higher loading in the evidence statements. However, for many areas of interest, systematic reviews were not available, so key well-referenced and structured narrative reviews were accepted if they concerned topics not covered elsewhere or if they added to better understanding. Policy papers, statements, and guidance documents were included where they demonstrated official positions or consensus.

Retrieved titles were screened, and abstracts obtained for all articles that appeared of possible interest. The abstracts were scrutinised by at least one of the reviewers to identify those of likely relevance, and discussed with the other to determine inclusion. Full papers were obtained by agreement. A broadly inclusive approach was used, so articles were included if they provided any material pertinent to the basic question 'Is work good for your health and well-being?', even if that was not the primary focus of the article. Exclusion was primarily on the basis of lack of relevance to that question. This open strategy for searching and selection was adopted to minimise the risk of failing to find or include articles that might contribute to the evidence synthesis.

A dedicated database was constructed, and all included papers were archived. Those articles deemed ineligible were excluded from the review, but retained for possible future use.

DATA EXTRACTION

Both reviewers appraised the included articles for information pertinent both to the overarching primary questions and to subsidiary issues under each topic. Data were extracted and entered into evidence tables variously by one of the reviewers and checked by the other; disagreements were resolved by discussion. For reviews, the nature of the review and the original authors' main conclusions are reported. For original studies, the characteristics are described and key findings reported. The present reviewers' comments on individual articles were added to the tables where appropriate.

EVIDENCE SYNTHESIS

Working together, the two reviewers progressively distilled and summarised the material from the evidence tables. Using an iterative process, evidence statements for each area were constructed, refined, and agreed. They were grouped by the review stages given in Table A1, and for ease of future referrence were given identifying letters based on the initial letter(s) of the heading concerned.

Each evidence statement was accompanied by a rating of the strength of the underlying evidence, with overt linkage to the evidence tables and supporting articles. The strength of the evidence for each evidence statement was rated on the scientific evidence using the definitions in Table A2.

	Scientific Evidence	Definition
***	**Strong**	generally consistent findings provided by (systematic review(s) of) multiple scientific studies.
**	**Moderate**	generally consistent findings provided by (review(s) of) fewer and/or methodologically weaker scientific studies.*
*	**Weak**	*Limited evidence* – provided by (review(s) of) a single scientific study. *Mixed or conflicting evidence* – inconsistent findings provided by (review(s) of) multiple scientific studies.
0	**Non-scientific**	legislation; practical, social or ethical considerations; guidance; general consensus.

TABLE A2: Evidence rating system used for the strength of the scientific evidence and evidence statements

[* scientifically weaker studies were those with methodological shortcomings such as low numbers, limited external validity, and possible bias]

The strength of the scientific evidence should be distinguished from the size of the effect: e.g. there may be strong evidence about a particular link between work and health, but the effect may be small. Where possible, effect sizes are indicated in the text of the evidence statements, but precise data on effect size was limited.

To enhance meaning, the text of the evidence statements was used to describe the character and contribution of the underlying evidence. Where the evidence statements were insufficient to convey complex underlying ideas, important issues were discussed in narrative text.

Finally, the entire material was progressively distilled into an evidence synthesis to reflect the overall balance of the evidence about work and health, together with any caveats and cautions. This was used to develop a a balanced model of the relationship between work and health in the context of healthy working lives.

QUALITY ASSURANCE

To guard against subjective bias in the search retrieval, data extraction, and evidence synthesis, a final draft of the report was peer reviewed by two internationally acknowledged experts in the field, with experience in occupational health; disability research, public policy, and epidemiology. They were asked to give their critical comments on the evidence statements and the strength ratings and synthesis, as well as to identify any evidence that may have been overlooked or misinterpreted. Their feedback guided refinement of the final report.